Environmental Education in the Anthropocene

Environmental Education in the Anthropocene explores how education can address the environmental crisis by combining theory, pedagogy, and practice. Organized into four sections, the book progresses from context to action: the foundations of the Anthropocene and its educational implications, the roles of institutions and actors in environmental education, innovative methodologies like experiential learning and transformative pedagogies, and practical tools for implementation. With a global perspective and local adaptability, the book offers case studies, strategies, and a roadmap for fostering sustainability education, empowering readers to drive change for future generations.

Features:

- Explains environmental education within the urgency of the Anthropocene to inspire action.
- Integrates global perspectives with locally adaptable solutions for diverse cultural and institutional contexts.
- Presents innovative pedagogical strategies, including participatory and transformative learning, for practical application.
- Offers real-life case studies and best practices to connect theory with actionable insights.
- Connects policy with practice to influence decision-making and drive sustainable education initiatives.

This book is intended for both academic and professional audiences, including instructors, graduate students, researchers, health and environmental professionals involved in projects related to environmental science, environmental chemistry, ecology, environmental pollution, and remediation.

Jorge Luis García Alcaraz is a professor at Universidad Autónoma de Ciudad Juárez, Mexico. He holds five PhD degrees in industrial engineering and related fields. A national researcher level three recognized by the Mexican National Council of Science and Technology, he specializes in lean manufacturing and advanced manufacturing technologies. He has authored and co-authored around 377 documents indexed in Scopus, including papers, conference papers, book chapters, and others. A founding member of the Mexican Society of Operation Research and a member of the Mexican Academy of Industrial Engineering, Dr García is also the author of several published books.

Ingrid Iovana Burgos Espinoza is an industrial engineer graduate from the National Technological Institute of Mexico. She holds a master's degree in industrial engineering and a PhD in Advanced Engineering Sciences from Universidad Autónoma de Ciudad Juárez, Mexico, with an emphasis on sustainability. Her research focuses on the impact of the attitudes and behavioral intentions of engineering students and active engineers and has shared her findings in international conferences. She has authored and co-authored several articles on environmental education and attitudes.

Environmental Education in the Anthropocene

Challenges and Solutions

Jorge Luis García Alcaraz and
Ingrid Iovana Burgos Espinoza

CRC Press
Taylor & Francis Group
Boca Raton London New York

CRC Press is an imprint of the
Taylor & Francis Group, an **informa** business

Designed cover image: Getty Images

First edition published 2027
by CRC Press
2385 NW Executive Center Drive, Suite 320, Boca Raton FL 33431

and by CRC Press
4 Park Square, Milton Park, Abingdon, Oxon, OX14 4RN

CRC Press is an imprint of Taylor & Francis Group, LLC

ISBN: 978-1-041-31552-0 (hbk)
ISBN: 978-1-041-31554-4 (pbk)
ISBN: 978-1-003-77966-7 (ebk)

DOI: 10.1201/9781003779667

Typeset in Times
by Apex CoVantage, LLC

Contents

SECTION II Actors and Ecosystem of Environmental Education

SECTION III Methodologies and Pedagogical Strategies

SECTION IV Implementation and Future of Environmental Education

Foreword

Over more than 26 years devoted to university teaching, engineering research, and collaboration with the industrial sector, I have had the privilege of observing—and at times contributing to—the development of researchers who, over time, have transcended the disciplinary boundaries of their initial training to address problems of truly global scope. This is neither a frequent nor an easy process. Few cases illustrate this evolution as clearly as that of Dr Jorge Luis García Alcaraz, whom I had the honor of supervising during his doctoral studies at the University of La Rioja and with whom I have since maintained a sustained and fruitful academic collaboration. It is particularly gratifying to observe how this vocation for training new researchers has been projected into his own work as a doctoral supervisor, as evidenced by his collaboration with Dr Ingrid Iovana Burgos Espinoza, co-author of this book, whose trajectory from industrial engineering to research in sustainability and environmental education reflects precisely the type of intellectual evolution demanded by contemporary challenges.

When Dr García Alcaraz and Dr Burgos Espinoza invited me to write the foreword to this work, I must admit that my initial reaction was one of intellectual curiosity. Both authors are trained in industrial engineering: He is widely recognized for his contributions to the optimization of production systems and supply chains, while she has a solid background in industrial safety and a strong focus on sustainability. However, after reading the 12 chapters organized into four sections, that initial curiosity evolved into a firm conviction: This work is not only relevant but also urgently necessary. Its strength lies in the perspective from which it has been conceived, situated at the intersection between the analytical rigor of engineering and the understanding of socioecological systems. In my view, this engineering approach enriched with social and ecological dimensions is precisely what is required in environmental education at this critical moment in human history.

Through these perspectives, the book demonstrates that the contemporary environmental crisis cannot be analyzed within traditional disciplinary boundaries but rather requires a genuinely holistic approach. Climate change, biodiversity loss, systemic pollution, and resource depletion must be addressed through multiple and simultaneous perspectives integrating natural sciences, engineering, social sciences, economics, ethics, and public policy. Consequently, any educational response that aims to be effective must reflect this complexity. It is here that the contribution of García Alcaraz and Burgos Espinoza becomes especially valuable: They have developed a work that, without oversimplifying complexity, offers the reader a coherent journey from global diagnosis to forward-looking proposals, integrating theoretical foundations, empirical evidence, institutional frameworks, methodological innovations, and documented case studies.

As an engineer, I have devoted my career to understanding how industrial systems can be optimized to become efficient, safe, and sustainable. In this context, the contribution of Dr Burgos Espinoza is particularly relevant, as her background in industrial safety and her research on the attitudes and behavioral intentions of engineers and engineering students toward environmental issues provide an empirical dimension that connects environmental education with professional behavior in real productive contexts. The combined experience of both authors reinforces a key idea present throughout the book: Technological transformation alone is insufficient. Many sophisticated technical solutions fail when they are not accompanied by a transformation in the awareness, values, and capabilities of the people who design, operate, and regulate these systems. Environmental education, as conceptualized in this work, serves as the bridge between technical capacity and transformative will—between what we know how to do and what we choose to do with that knowledge.

A notable strength of this work is its solid grounding in up-to-date scientific evidence. The authors present a review of contemporary literature, with particular emphasis on studies published since 2020 while also acknowledging the seminal contributions that have shaped the field. In a domain evolving as rapidly as environmental education, having access to a rigorous and updated synthesis of current knowledge is of immense value for researchers, educators, and decision-makers.

The structure of the book follows a progressive argumentative sequence that reflects the intellectual maturity of its authors. It begins with a multidimensional analysis of the planetary crisis and moves toward the examination of challenges and future perspectives, addressing stakeholders, methodologies, institutional frameworks, and scaling strategies. Each chapter builds upon the previous ones, preparing the ground for those that follow. This organization facilitates both sequential reading and selective consultation, making the book a valuable reference for audiences with diverse interests and needs.

I would also like to highlight the deeply human dimension underlying this work. I have known Dr García Alcaraz personally for over a decade and witnessed his commitment to sustainability—not as an abstract academic stance but as a conviction rooted in his experience as a researcher, as a professional engaged with industry, and, above all, as a father. Dr Burgos Espinoza shares this conviction through her own trajectory: Her transition from industrial engineering to research in sustainability, through the Doctoral Program in Advanced Engineering Sciences at the Autonomous University of Ciudad Juárez, reflects a genuine commitment to finding educational responses to contemporary environmental challenges. This shared authenticity is evident in the tone of the book: Rigorous without being cold, comprehensive without being tedious, and committed without being naive. The synergy between both authors—he contributing a broad vision shaped by years of experience between academia and industry, and she bringing a fresh perspective grounded in the study of new generations—results in a work of coherence, depth, and relevance.

In the Ibero-American context, where universities face the dual challenge of generating cutting-edge knowledge while contributing to the social development of their communities, this work takes on particular significance. The Autonomous University of Ciudad Juárez, the institutional home of both authors, operates in a border region where tensions between industrial development, social justice, and environmental protection are especially pronounced. It is from this concrete reality—not from an ivory tower—that this book has emerged, granting its analyses an authenticity and urgency that transcend purely academic exercise.

I invite the reader to approach this work not as an encyclopedic compilation for occasional consultation but as an intellectual map that guides action. Each chapter offers not only diagnoses and analyses but also practical guidance, evidence of what works, and reflections on the paths ahead. At a time when humanity faces environmental decisions whose consequences will extend across generations, having educational tools grounded in the best available evidence is not an academic luxury, but a civilizational necessity.

The magnitude of the environmental challenges we face can be overwhelming. However, as the authors convincingly demonstrate throughout these pages, environmental education—when designed with rigor, implemented with commitment, and evaluated with honesty—has the capacity to transform not only what people know but also what they value, what they do, and ultimately the kind of future they are willing to build. This book is, in itself, an act of environmental education: it does not merely describe problems but helps shape the judgment required to address them. It reminds us that hope, when grounded in evidence and action, ceases to be a mere aspiration and becomes a form of intergenerational responsibility.

Emilio Jiménez Macías, PhD
Full Prof. of Systems Engineering and Automation.
University of La Rioja
Logroño, Spain
March 2026

Preface

This book is born from the convergence of two trajectories that have defined our professional lives by working as academics in public education institutions and in industry, simultaneously. As academics, we have seen through scientific research that environmental aspects have gained interest, and as professionals in the industry, we have been in direct contact with the production systems in our environment. Development in the educational and industrial sectors has allowed us to have a complementary perspective: One in which we read and investigate environmental degradation through scientific studies, which converges with the environmental metrics of production systems and their supply chains.

This double professional performance has allowed us to materialize this work. We understand that the environmental crisis that humanity is currently facing is not only a technical problem but also an educational challenge that allows us to profoundly transform the way in which we, as individuals, communities, and institutions, understand, value, and interact with nature. Currently, scientific studies show that the limits of our planet have been exceeded, as biodiversity is declining rapidly, climate systems are being destabilized, and these changes will redefine human existence and the conditions in which it will occur in the future. However, to act requires translating these metrics into behavioral and political changes that allow a transformation, and it is still incomplete. In this sense, environmental education (EE) is an ambitious approach that can be the critical bridge between what we know and what we do.

However, this book also results from the personal concern that we authors have. As human beings who are integrated into a family and a society, we are aware today of the need to have better environmental decisions, since the actions we take today will be reflected as an ecological inheritance that we leave to our children and grandchildren. We assume intergenerational equity, where they also have the right to have availability of resources and a clean planet. We clearly understand that the future of our children depends on the decisions, policies, and technologies that we develop, so it is urgent to work on creating better environmental awareness.

This book is organized into 12 sequential chapters, analyzing the topic from diagnosis to action and integrating basic concepts with a vision for the future. In this book, the authors assume that better decision-making requires an adequate understanding of the environmental crisis and the capacities that can be achieved through EE.

Chapter 1 analyzes the current environmental crisis from a multidimensional approach that includes climate change, biodiversity loss, pollution, and the accelerated depletion of natural resources, thus justifying the need to implement EE programs. Chapter 2 presents some of the conceptual and theoretical foundations that are basic in EE, where the origin of EE, the study of nature and its conservation, and the regulatory frameworks that constitute what we now call sustainability, including transformative learning and the models that generate it, are analyzed.

Chapters 3, 4, and 5 analyze the impacts obtained from EE, as well as the modalities and scenarios in which it is generated. Specifically, Chapter 3 examines the transformative effects generated by EE, which are related to knowledge, attitudes, values, and behaviors as individuals. Chapter 4 explores the formal and institutional pathways through which EE is obtained, such as colleges and universities. Chapter 5 examines the nonformal and community environments through which EE is obtained, where corporate responsibility of companies is integrated.

Chapter 6 is vital because it analyzes the policies, governance, and institutionalization that facilitate or limit EE at the local, national, and international levels. Chapter 7 focuses on innovative methodologies that enable environmental knowledge and learning, emphasizing branching and current digital technologies that are traditionally used by environmental educators. Chapter 8 analyzes curricular design and adaptation and discusses how the incorporation of knowledge based on traditional logic facilitates people's professional performance.

Chapters 9 and 10 discuss the quality and scope of EE programs. Specifically, Chapter 9 discusses assessment methodologies and the myriad challenges faced in measuring multidimensional learning of EE concepts. Chapter 10 discusses strategies for scaling and implementing EE programs in different settings. Chapter 11 offers a series of success stories and good practices documented in different geographical settings and educational institutions, highlighting how EE can be effective when designed holistically.

Finally, Chapter 12 reports on many of the challenges currently faced and proposes some future perspectives in which concepts and tools related to artificial intelligence and emerging and contemporary technologies are integrated; in addition, it reports a reconceptualization of EE for the Anthropocene era and the development of citizenship as an educational aspiration for the period 2030–2050.

WHAT APPROACH DO WE USE IN THIS BOOK?

Readers of this book should understand how it was generated. The authors wish to state that this book is the product of a review of updated scientific literature, focusing mainly on articles that have already passed through a peer-review process. However, we wish to state that this temporal approach does not exclude seminal contributions to the topic of EE in previous decades. Orienting ourselves with recent information has allowed us to understand the emerging trends and innovative approaches that continue to be generated with great dynamism. These old works, when they are cited, it is because they are very solid theoretical foundations, empirical studies that have been emblematic or that constitute conceptual frameworks that continue to have an influence on current research and practice.

This decision to focus on recent literature is also due to the fact that the contexts in which EE is applied change rapidly because of technological and social advances and high ecological impact. Likewise, the acceleration of environmental crises, the emergence of transformative digital technologies, and the lessons we have gained from observing global disruptions, such as COVID-19, have contributed to a dizzying evolution in which EE research and practice must adapt to current environments.

WHO SHOULD READ OUR BOOK?

This book is aimed at a broad academic and professional audience, including researchers in EE and sustainability sciences, educators and curriculum designers at all levels, decision-makers responsible for educational and environmental governance, institutional leaders seeking to integrate sustainability into organizational practice, and graduate students pursuing advanced studies in EE or related fields. Although the treatment is academic and evidence-based throughout, the practical implications of the reviewed research are consistently highlighted to ensure relevance for both practitioners and theorists.

Jorge Luis García Alcaraz, PhD
Ingrid Iovana Burgos Espinoza, PhD
Universidad Autónoma de Ciudad Juárez
Ciudad Juárez, Chihuahua, México
March 2026

Since the authors are not native English speakers, they would like to acknowledge the use of Grammarly Pro and Paperpal PRIME for proofreading purposes and to help improve the readability of the chapters.

Acknowledgments

We express our deepest gratitude to the National Council of Humanities, Sciences, and Technologies (CONAHCYT) of Mexico and the National System of Researchers (SNII) for their institutional support, which has sustained our research programs over the years. Likewise, we gratefully recognize the academic environment provided by the Autonomous University of Ciudad Juárez, our institution of affiliation, which has provided us with the intellectual freedom and collegial support necessary for a project of this magnitude.

We also thank the editorial team of CRC Press for their professionalism and support throughout the production of this work.

Finally, and most fundamentally, we dedicate this book to our children and to all the children who will inherit the consequences of the environmental choices made by the present generation. We hope that the knowledge synthesized in these pages can contribute, however modestly, to building the educational foundations on which a more sustainable, equitable, and ecologically conscious world can be built. The magnitude of the environmental challenges we face demands nothing less than the full mobilization of our educational capacities in service of the future.

Jorge Luis García Alcaraz, PhD
Ingrid Iovana Burgos Espinoza, PhD
Universidad Autónoma de Ciudad Juárez
Ciudad Juárez, Chihuahua, México
March 2026

Section I

Foundations of the Environmental Crisis and Education

1 The Planet in Crisis
Diagnosis and Urgency

1.1 GLOBAL OVERVIEW OF THE ENVIRONMENTAL CRISIS

The world's population is at a crucial juncture. Natural systems that have sustained life on Earth for so long are showing signs of rapid deterioration. Many scientists believe that this is the sixth mass extinction and the beginning of the Anthropocene epoch (Ceballos et al., 2015; Crutzen, 2002). This major environmental crisis is of great importance to humans worldwide. This is evident every day, everywhere, in the form of extreme weather events. This is evident in the loss of animals and plants, pollution, and the excessive use of natural resources.

Human-caused climate change has become the most obvious and urgent environmental problem worldwide. The amount of carbon dioxide in the air has exceeded 420 parts per million, a level not seen for millions of years (Change, 2023). This increase, mainly due to the burning of fuels such as oil, is significantly changing the climate, with negative effects on both nature and people. Glaciers are melting at an excessively rapid rate; as a result, sea levels are rising, endangering coastal cities, and severe weather events such as hurricanes, droughts, and floods are becoming more frequent (Change, 2023).

Biodiversity loss is another important issue in the current environmental crisis that requires urgent attention. Experts estimate that species are now disappearing at a rate a thousand times higher than the normal rate of extinction, creating the sixth mass extinction in Earth's history (Pimm et al., 2014). Forest ecosystems, wetlands, coral reefs, and seagrass beds are rapidly deteriorating, jeopardizing key ecosystem services that support human well-being. Deforestation continues unabated, especially in tropical areas such as the Amazon, where millions of hectares of forests are removed each year.

Pollution on the planet is enormous and damages air, water, land, and food quality. Microplastics have been found in the deepest seas and highest mountains (Jamieson et al., 2019). This indicates that human-caused pollution is ubiquitous. Some synthetic chemicals persist for long periods and accumulate in the bodies of living organisms. These chemicals travel through the environment, disrupting important processes and endangering the health of animals and plants.

According to a study by Babintseva et al. (2025), it is very important to teach about sustainable development. This is because "environmental education helps to create a responsible, humane, and correct way of thinking about nature." As environmental problems are global, education seeks to change the way we think and act in relation to nature.

Environmental crises are complex and interconnected. Problems are not isolated but are interrelated. This can cause chain reactions and irreversible changes in the

DOI: 10.1201/9781003779667-2

Earth's systems. Global warming is melting Arctic ice faster, releasing methane and carbon dioxide into the atmosphere (Schuur et al., 2015). This exacerbates the warming. Ocean acidification, caused by CO_2 absorption, damages marine ecosystems and the fishing industry, on which millions of people depend (Doney et al., 2009).

Safardokkht bahar et al. (2025) used a meta-synthesis in environmental training based on carbon footprint management in gas refineries shows how difficult it is to address specific environmental problems. Their analysis emphasizes that "the gas refining industry, as one of the most important industries and areas of economic creation in Iran, influences the global economy," but that economic considerations should not overshadow the environmental effects of these industries, including greenhouse gas emissions, which are one of the most important environmental problems.

The duration of the environmental problem adds urgency to this situation. The modes of environmental deterioration range from rapid changes to alterations that take years or even centuries to fully manifest. This lack of temporal synchrony between causes and effects creates unique challenges for how people view this and for policy responses. Today's greenhouse gas emissions will define climate conditions for many years to come, whereas the effects of chemical pollution may be observed for many generations after first contact.

The uneven geographical distribution of environmental impacts is an aspect of environmental justice that makes the global context more challenging. The most economically and socially vulnerable groups almost always face the most severe environmental impact (Diffenbaugh & Burke, 2019). In contrast, those who contribute most to environmental degradation may be somewhat spared from its direct consequences. This environmental inequality is evident at both regional and global levels, where developing nations suffer disproportionately from the effects of climate change despite contributing very little to past emissions.

1.2 MAIN TYPES OF ENVIRONMENTAL POLLUTION (AIR, WATER, SOIL, NOISE)

1.2.1 Air Pollution: An Invisible But Omnipresent Threat

Air pollution is a major threat to health and the environment worldwide. The World Health Organization states that pollution causes more than seven million deaths worldwide annually. This makes it one of the biggest health problems worldwide. A large number of studies showed that many people in cities and towns were exposed to dangerous levels of air pollution.

The main types of air pollution include particulate matter, nitrogen and sulfur oxides, carbon monoxide, and volatile organic compounds (VOCs). These emissions originate from vehicles, factories, power plants, and agricultural activities. Very small fine particles are hazardous. They can enter the lungs and reach the bloodstream, causing problems in the heart, respiratory system, and brain (Pope III et al., 2002).

Other types of pollution are formed when primary pollutants mix with the air. Tropospheric ozone, a component of smog, is formed when nitrogen oxides and VOCs react with sunlight. Although stratospheric ozone protects us from ultraviolet radiation, ground-level ozone is harmful because it harms our lungs and plants.

Cruz-Quijada and Tejedor-Flores (2024) analyzed environmental education in solid waste management in Panama. This study shows important links between the different types of pollution. This descriptive study found that "important steps have been taken across the country to promote environmental education and recycling. However, policies and curricula for environmental education still need to be developed in the country. This would help teachers, public institutions, and NGOs to manage the complexity of solid waste management." Vehicles are a major source of air pollution in urban settings. Car engines emit several pollutants, including nitrogen oxides, diesel particles, hydrocarbons, and heavy metals. The rapid increase in car ownership worldwide, particularly in developing countries, has exacerbated these problems (Karagulian et al., 2015). This is despite improvements in the technology used to control the emissions.

Industries also heavily pollute the air. Power plants, refineries, and steel mills emit numerous pollutants. These can travel long distances and affect entire regions. Industries emit common pollutants and toxic substances such as mercury, lead, dioxins, and organic compounds. These pollutants can accumulate in food chains.

1.2.2 Water Pollution: Threatening the Most Vital Resource

Water, which is vital to life, is heavily polluted, reducing its quality and availability for humans, animals, and agriculture. Water pollution is observed in rivers, lakes, seas, and groundwater, which is crucial for the future of humanity. The connection between water pollution and inadequate waste management underscores the need for comprehensive educational approaches. Ferronato and Torretta (2019) conducted a systematic review of waste mismanagement in developing countries and found that "inadequate waste management practices, including open dumping and uncontrolled landfilling, are major contributors to the contamination of surface water and groundwater in low- and middle-income countries." Their analysis highlights that educational interventions targeting waste reduction, separation at source, and community engagement are essential components of integrated water pollution prevention strategies.

There are many different types of factory chemicals, and they often end up in water sources. Metals such as mercury, lead, cadmium, and chromium can remain at the bottom of the food chain for years, where aquatic animals absorb and accumulate them in their bodies. Synthetic organic compounds, such as pesticides and medicines, are hazardous because they affect living organisms and can disrupt the endocrine system, even at low concentrations.

Nutrient pollution, such as nitrogen and phosphorus from fertilizers and sewage, causes excessive algal growth in aquatic ecosystems (Diaz & Rosenberg, 2008). When these algae die, they consume oxygen in the water, creating lifeless zones. The Gulf of Mexico, the Baltic Sea, and many lakes clearly show the effects of human-caused pollution on the environment.

Microplastics are a widespread pollutant (Thompson et al., 2004). They affect everything, from oceans to the water we drink. These tiny particles originate from large pieces of plastic waste that are broken down. They also originate from synthetic fibers in clothing and microbeads in cosmetics products. Their small size makes them prey to aquatic animals. This occurs at all levels of the food chain in the ocean. The harmful effects of this on the environment are still under investigation.

1.2.3 Soil Pollution: Degrading the Basis of Life on Earth

Soil, often referred to as the "living skin" of the Earth, is a complex ecosystem that aids in food production, regulates water cycles, stores carbon, and supports a wide variety of life forms. Soil pollution jeopardizes these important tasks, threatening food security and the stability of land ecosystems.

Pollutants that contaminate soil include heavy metals, which can persist for extended periods and alter microbial communities integral to natural processes. Lead, which comes from years of use in gasoline and paint, contaminates soils in and around cities, where it harms children's brain development. Mercury, which originates from mining and coal burning, accumulates in forest soils and affects all ecosystems adversely.

Pesticides and herbicides, widely used in agriculture, can persist in the soil for years, affecting non-target organisms such as pollinators, natural predators, and soil microbes. Some organochlorine pesticides, although no longer permitted in many countries, remain in the soil because they are highly stable and can persist for many years after their use (Jones & de Voogt, 1999).

Dos Santos Castro and Soares (2024) report a paper entitled "Green market: a tool for recycling and environmental education" and underscore the importance of addressing soil contamination through multifaceted approaches. Their study states that "the activity of collecting recyclable waste in the riverside community of Catalao, in the metropolitan region of Manaus, served to emphasize the importance of selective collection in transforming waste into a sustainable economy within the confines of the circular economy."

Hydrocarbons from petroleum contaminate soil through accidental spills, leaks from underground storage tanks, and improper waste disposal. These compounds can form clouds of pollution that spread laterally and downward, affecting groundwater and posing health hazards to humans through evaporation and direct contact with the soil.

1.2.4 Noise Pollution: The Invisible Pollutant

Noise pollution, which is sometimes not considered as important as other types of pollution, is increasingly affecting both cities and the countryside. It also changes how animals live and survive in their habitats. Human-generated noise has increased significantly in recent years. This has altered natural sounds and created noisy environments in which many species cannot survive. Noise originates from cars, factories, construction sites, airplanes, and leisure activities. In cities, automobiles are the main source of noise pollution. This noise is often louder than that recommended for good health. Car engines, brakes, and wheels produce different types of noises. These noises vary depending on the type of car, its speed, and the road conditions.

Prolonged exposure to noise can cause sleep disturbance, cardiovascular strain, and cognitive impairment (Münzel et al., 2014). This is because noise activates the nervous system and triggers the release of stress hormones. If these hormones are elevated for a long time, they can cause high blood pressure, heart problems, and metabolic disorders.

For animals, noise interferes with their ability to communicate, which is important for reproduction, foraging, and predator avoidance. Birds, which use many sounds, change their songs depending on where they live and how many young they have in noisy places (Slabbekoorn & Peet, 2003). Likewise, marine mammals, such as whales, which use sound to locate and communicate with each other, face numerous challenges owing to noise from ships, oil exploration, and construction (Hildebrand, 2009).

1.3 PRIMARY AND SECONDARY SOURCES OF POLLUTION

1.3.1 Basic Classification of Pollution Sources

Understanding the causes of pollution and its mechanisms is key to developing plans to prevent, control, and address environmental issues. Pollution sources can be divided into primary and secondary sources. Primary sources directly release pollutants, whereas secondary sources are formed through physical, chemical, or biological transformations in the environment.

Primary sources of pollution release pollutants into the environment, including activities that emit harmful substances. For example, cars that emit carbon monoxide and nitrogen oxides into the air, factories that discharge heavy metals into water, and waste that can leach toxic substances into soil and groundwater.

Secondary sources create pollution by transforming substances. One example is tropospheric ozone, which is not directly released into the atmosphere. It is formed when nitrogen oxides and VOCs react in the presence of heat and sunlight. This is important because reducing secondary pollution requires the management of multiple factors and accounting for environmental conditions that contribute to its formation.

The study by Persano Adorno et al. (2025), titled "Sustainability as a Cross-Curricular Link: Creative European Strategies for Eco-Conscious Environmental Education," offers important ideas for managing various sources of environmental problems. Their Erasmus+ project, "clean environment-clean school climate with creative environmental practices in school education-clean & creative" aims to develop thematic environmental curriculum content that integrates seamlessly across ten STEAM disciplines.

1.3.2 Point Sources versus Diffuse Sources

The location of emissions is another key criterion for classifying them, as it affects their monitoring, regulation, and control. Point sources release pollution from specific, clearly identifiable locations, allowing direct measurements and precise control techniques to be applied at the point of release.

Factory smokestacks, tailpipes, and landfills are point sources that can be monitored and regulated using specific permits and emission limits to reduce air pollution. Diffuse or non-point sources release pollution from large areas without discrete points of discharge, making them much more difficult to monitor and control than point sources. Rainwater that carries fertilizers and pesticides from thousands of fields into rivers and lakes is a great example of diffuse or "non-point" pollution. City exhaust is another form of diffuse pollution, as many cars emit exhaust and collectively cause significant harm to the environment.

Deposition in the air shows how pollution from specific sources can spread widely, acting as diffuse sources for terrestrial and aquatic life that are far from the source. Mercury released by power plants can travel thousands of kilometers through the air before falling into distant rivers and lakes, where it mixes and accumulates in the food chain.

1.3.3 Industrial Activities as Sources of Pollution

Factory activities are a major source of environmental pollution. This varies substantially across industries, machine types, materials processed, and pollution control practices. Many hazardous substances are used in the chemical, oil, and pharmaceutical industries. This produces air emissions, dirty liquids, and solid waste that require special care and disposal methods.

Industries that extract materials from the earth, such as mining and oil extraction, cause significant environmental harm. This is because they are large-scale operations that use hazardous substances. Open-pit mining significantly alters landscapes. It also exposes minerals that can generate acidic liquids for many years and releases heavy metals that contaminate the soil and water.

Oil refineries, as shown in the study by Safardokkht bahar et al. (2025), are complex sources of pollution that require comprehensive approaches to environmental protection. Their study on controlling the carbon footprint in gas refineries states that "measuring and evaluating greenhouse gas emissions requires learning how to manage the carbon footprint, which is very important in this area." The clothing industry causes significant water pollution during the dyeing and fabric finishing processes. These processes use synthetic dyes, heavy metals, and chemicals. Many of these are long-lasting, accumulate in living organisms, and are toxic, causing health and environmental problems in areas with high clothing production.

1.3.4 Agricultural and Livestock Sources

Despite being key to global food security, agriculture is a major source of environmental pollution through the intensive use of synthetic fertilizers and pesticides, as well as through management practices that can degrade soil, contaminate water, and harm biodiversity. The use of nitrogen fertilizers can lead to nitrate leaching into groundwater and runoff, thereby contributing to the eutrophication of surface waters.

Pesticides used in agriculture can poison various environmental components. Movement during use can affect nearby crops, local vegetation, and wildlife habitats. The persistence of some pesticides in the soil can lead to continuous exposure of soil organisms and contamination of subsequent crops. Evaporation can transport pesticides to distant areas where they can accumulate and enter vulnerable ecosystems.

Intensive livestock farming causes pollution in many ways, including methane and nitrous oxide emissions that contribute to climate change, the generation of waste with high levels of organic matter and nutrients that can contaminate surface and groundwater, and ammonia emissions that contribute to the acidification and eutrophication of ecosystems.

1.3.5 Urban and Domestic Sources

Although activities in cities and homes are small, they are major sources of pollution when considered collectively, given the large number of people and activities in urban areas. Wastewater from homes contains harmful microbes, food, medicines, and toiletries that may persist even after conventional treatment.

Solid waste from cities is a challenging source of pollution because it contains materials that degrade naturally and produce liquids and gases. Some items can be reused, but if improperly disposed of, they can pollute land and water, along with hazardous waste such as batteries, appliances, and cleaning products.

Transportation in cities contributes significantly to air and noise pollution problems. Cars emit not only gases but also particles created by the wear and tear of brakes, tires, and the ground. It is also important to note that the construction of transportation routes can fragment natural habitats and create barriers to wildlife movement, thereby creating new problems for wildlife.

1.3.6 Chemical Transformations and Secondary Processes

Secondary processes that convert primary pollutants into potentially more dangerous products add significant complexity to environmental quality management. Photochemical reactions in the atmosphere can transform relatively harmless hydrocarbons into powerful and highly reactive oxidants. Methylation processes convert inorganic mercury into methylmercury, an extremely neurotoxic form that accumulates efficiently in aquatic food webs.

Biological degradation methods can produce metabolites that are more toxic than the original compounds. Some organophosphate pesticides break down into products that retain or increase their toxicity. The incomplete biodegradation of complex organic compounds can generate reactive intermediates that may be more problematic than their parent compounds.

Ding et al. (2025) report that a paper entitled "Facing Challenges in Collaborative Environmental Education On and Off Campus" exemplifies the complexity of addressing diverse sources of environmental problems. Their research on the primary and secondary education project examined the current state and challenges of collaboration in environmental education, both on and off campus, from teachers' perspectives, identified key issues, and suggested strategies for improvement.

1.4 INTERCONNECTION BETWEEN LOCAL AND GLOBAL ENVIRONMENTAL PROBLEMS

1.4.1 The Systemic Nature of Environmental Problems

Today's environmental challenges span geographical, political, and temporal boundaries and manifest as complex systems in which changes in one area can have global repercussions. This essential interrelationship demands approaches to understanding that go beyond fragmented views to embrace the integrated nature of human–environment interactions during the Anthropocene.

Climate change is a clear example of the connection between local and global levels. Greenhouse gas emissions from specific local activities spread into the global atmosphere, contributing to changes in climate patterns that manifest differently across regions.

A power plant in one country can influence global warming, which could exacerbate droughts in one area, cause flooding in another, and accelerate glacier melting in yet another, thereby generating a chain of effects that ultimately affect ecosystems and communities globally.

Hemby et al. (2024) demonstrated how local access problems in environmental education reflect broader trends in the accessibility and distribution of environmental education programs through field trips for US adolescents in the United States. Their nationwide research revealed that geographic access was concentrated in certain areas, especially in diverse regions, and that spatial access was strongly linked to political leanings and the level of urbanization, with more rural, White, and Republican-leaning areas generally having significantly less access to environmental education.

1.4.2 Global Atmospheric Transport Systems

The atmosphere acts as a global transport system, moving pollutants released in one place to distant areas. Environmental problems are often transferred from high-emission areas to ecosystems and community groups that contribute little to the initial pollution. This movement in the atmosphere occurs at various spatial and temporal scales, ranging from the regional transport of pollutants over a matter of days to global circulation that can extend over weeks to months.

Persistent organic pollutants are a clear example of these global transport phenomena in the environment. Chemicals such as dichlorodiphenyltrichloroethane (DDT) and polychlorinated biphenyls (PCB), which are mainly emitted in industrial and agricultural areas, are transported by air to polar regions, where they condense because of low temperatures. This causes them to accumulate in Arctic ecosystems thousands of kilometers from their original locations. This phenomenon, known as global distillation, has led to significant contamination of Arctic wildlife and exposed indigenous communities that depend on marine resources.

Mercury emitted by power plants, mining activities, and industrial processes is transported through the atmosphere in its elemental form (Driscoll et al., 2013). It is then oxidized and deposited in aquatic environments, where it is transformed into methylmercury, a highly toxic compound that accumulates in the food chain. This process links local mercury emissions to fish contamination in distant lakes, posing health risks to communities that depend on fishing for their livelihood.

1.4.3 Transboundary Water Interconnections

Aquatic systems link regions and ecosystems through watersheds that often cross political boundaries, establishing dependencies in which conditions in the upper parts of a watershed can significantly influence environmental conditions and resource availability in lower areas. These interconnections transport not only water but also sediments, nutrients, pollutants, and aquatic life, thereby forming complex ecological and social relationships.

Eutrophication in aquatic systems exemplifies this relationship. Fertilizers used in agricultural practices in the upper reaches of watersheds can leach into water bodies, reaching lakes, reservoirs, and coastlines, where they promote excessive algal growth, reduce dissolved oxygen, and create dead zones. The Gulf of Mexico receives nutrients from the Mississippi River Basin, which drains agricultural areas in several states, forming a dead zone that harms commercial fisheries and marine ecosystems (Rabalais et al., 2002).

Transboundary watersheds require international collaboration to manage water resources and effectively combat the pollution. Rivers such as the Danube, Nile, and Mekong flow through several countries, where differing environmental policies, levels of development, and economic priorities can lead to disputes over water use, infrastructure construction, and pollution management.

1.4.4 Socioeconomic Telecouplings

Telecouplings are economic and social interactions between distant systems that are interconnected through the exchange of information, energy resources, materials, living beings, and human resources (Liu et al., 2013). These relationships can propagate environmental effects over long distances, creating indirect ecological responsibilities and weakening the link between consumption habits and their environmental impact.

Global trade in agricultural and forestry products clearly illustrates these telecouplings. The increased demand for soybeans in developed nations has encouraged agricultural expansion in tropical regions, leading to deforestation in the Amazon and Cerrado regions of Brazil. Although soybean consumers may be physically distant from these environmental effects, their purchasing choices indirectly contribute to them.

Tourism is a powerful example of socioeconomic telecoupling, as the consumption choices of travelers in one region can drive environmental pressures in distant destinations. Ardoin et al. (2015) conducted a comprehensive review and analysis of the literature on nature-based tourism's impact on environmental knowledge, attitudes, and behavior. Their findings indicate that "nature-based tourism experiences can positively influence visitors' environmental knowledge and attitudes, and in some cases, their environmental behaviors," although the durability and depth of these changes vary depending on the educational design of the experience. This research underscores the potential of environmental education programs embedded in tourism to bridge the gap between distant consumers and local environmental impacts, fostering a sense of ecological responsibility that transcends geographical boundaries.

1.4.5 Ecological Cascades and Domino Effects

Natural environments are linked by complex networks of biological interactions that can transmit disturbances across trophic levels and habitats, generating cascading effects that intensify local impacts on regional and global scales. These ecological chains can operate through various mechanisms, including competition, mutualism, and habitat change.

The disappearance of top predators can trigger cascades that alter community structure and ecosystem function across large spatial scales. The eradication of wolves in Yellowstone National Park has altered the behavior of herbivores and affected plant diversity (Ripple & Beschta, 2012). It alters the structure of riparian ecosystems, demonstrating how the removal of a crucial species can generate ecological effects that spread across multiple components of the ecosystem.

Habitat division can cause edge effects that extend into the remaining fragments, altering the microclimate, species diversity, and ecological processes. These effects can spread across fragmented landscapes, linking local alterations to regional changes in biodiversity and ecosystem function.

1.4.6 Species Migration and Dispersal

The movement of organisms, whether driven by natural forces or human activities, links distant ecosystems and can spread ecological effects over large areas. Migrating species connect breeding, feeding, and hibernation areas that may be thousands of miles apart, and transport nutrients, energy, and pollutants between different ecosystems.

Non-native species demonstrate how local disturbances can have global repercussions through organism movement. The introduction of species into new habitats, whether accidental or deliberate, can lead to invasions that transform the original ecosystems, affect native species, and alter the ecosystem functions. Zebra mussels, which were introduced into the Great Lakes of North America, have spread widely, altering aquatic ecosystems and causing significant economic impact (Karatayev et al., 2015).

1.4.7 Global Climate Feedback

The Earth's climate system operates through complex interactions that can amplify or dampen disturbances locally, generating nonlinear patterns in which minor changes can have disproportionate effects on the climate. These interactions link local activities to global climate trends, potentially leading to drastic transformations and abrupt shifts.

Arctic ice melt is a clear example of positive feedback that increases the rate of global warming. The decline in sea ice reduces the albedo of the Arctic region, leading to greater absorption of solar radiation, which, in turn, accelerates warming and further melting of the sea ice. This interaction links local physical actions in the Arctic to global climate change.

The release of carbon from permafrost is an alarming climate feedback mechanism. Rising global temperatures are melting permafrost that has remained frozen for thousands of years, releasing carbon dioxide and methane, which intensifies global warming (Turetsky et al., 2020). This feedback can establish climate commitments in which future warming depends on processes that have already begun.

1.4.8 Systemic Vulnerabilities and Tipping Points

The connection between local and global environmental problems creates vulnerabilities within systems, where seemingly insignificant alterations can spread through complex structures, potentially causing their collapse or radical changes. These vulnerabilities within the system are particularly alarming because they can lead to irreversible transformations that permanently alter the functioning of socioecological systems (SES).

Climate thresholds are crucial limits at which small additional changes can cause abrupt or irreversible changes in climate systems. Glacier collapse, changes in ocean currents, and tropical rainforest deforestation are examples of thresholds that can significantly alter global climate (Lenton et al., 2008). The study conducted by Lozano-Lucia and Talavera-Ortega (2024) and entitled "Environmental education methodology in childhood: a comparison between educators in Spain and Costa Rica" highlights the importance of addressing these complex interrelationships from an early age. According to their research, although there are some notable differences, there is, in general, a shared global vision across the topics addressed, underscoring the need to improve educators' initial training.

1.5 CONCLUSION

The current environmental crisis is the most complex and urgent challenge facing humanity, characterized by systemic interconnections across geographical, temporal, and disciplinary domains. Building on this chapter, the following analysis shows that environmental problems have arisen from a profoundly altered global system in which human activity has affected fundamental planetary processes.

The scale and speed of contemporary environmental changes define a new geological epoch, the Anthropocene, during which humans have emerged as the dominant force shaping Earth's systems. This situation requires a paradigm shift in how we understand the interactions between society and nature, moving beyond conventional binaries to develop a more integrated view of complex SES (Rockström et al., 2009).

The various forms of pollution analyzed, including air, water, soil, and noise, illustrate the centrality of anthropogenic disturbances at all levels of the environment. The systemic nature of these problems requires comprehensive plans to address the environmental crisis by recognizing the interconnections among different types of pollution and their various sources. The distinction between primary and secondary pollution sources is critical because implementing robust preventive and control measures is difficult without it.

The chemical and biological processes that convert primary pollutants into harmful secondary products, can only be described in advanced terms so that it is easy to appreciate and determine management strategies that not only are more comprehensive but also consider the entire life cycle of pollutants in environmental processes. However, the links between local and global environments highlight that the environmental crisis is too complex to be resolved by pursuing a series of separate measures or fragmented policy actions.

Socioeconomic telecoupling, global atmospheric transport, transboundary water connections, and cascades of ecological dynamics create interdependent chains that require global collaboration and transnational governance approaches that function effectively at all scales. The evidence for the above points is the way forward.

Environmental education is a fundamental strategy for addressing global environmental crises. Environmental education can strongly influence environmental knowledge, attitudes toward the natural environment, and related behaviors, as documented in several studies (Ardoin et al., 2018). However, the effectiveness of environmental education critically depends on its ability to address the systemic complexities of contemporary problems.

Educational approaches that view environmental problems as merely isolated issues or that focus only on individual behavioral changes may, in fact, be inadequate to achieve the systemic transformations required by the scale of the current crisis. The urgency of the environmental crisis necessitates effective approaches to educate students and staff about it. The windows of time to avoid irreversible climate or environmental tipping points are closing rapidly, and educational interventions can produce social impacts on an unprecedented scale and at unprecedented speeds.

The argument also holds that environmental crises are closely linked to social justice and equity issues. The most vulnerable populations bear the brunt of environmental damage and contribute the least to environmental degradation, whereas the opposite is true for the least vulnerable populations. This aspect of environmental justice must be integral to any educational response that seeks to address the root cause of environmental crises.

Finally, the global nature of the environmental crisis dictates that environmental education cultivates the capacity for global citizenship. This capacity recognizes responsibilities and potential that national boundaries cannot define. This includes fostering awareness of the interconnectedness of the planet's many systems, empathy for communities and ecosystems elsewhere, and a commitment to international cooperation to address shared environmental challenges.

Now that you are familiar with the environmental crisis documented in this chapter—which provides an important backdrop for understanding why environmental education is critical to the future of humanity and the planet—the following chapters will explore at how ecological education can respond to this crisis, exploring its theoretical foundations and historical development, and how this could be an approach to social transformation toward a sustainable future.

REFERENCES

Ardoin, N. M., Bowers, A. W., Roth, N. W., & Holthuis, N. (2018). Environmental education and K-12 student outcomes: A review and analysis of research. *The Journal of Environmental Education*, *49*(1), 1–17. https://doi.org/10.1080/00958964.2017.1366155

Ardoin, N. M., Wheaton, M., Bowers, A. W., Hunt, C. A., & Durham, W. H. (2015). Nature-based tourism's impact on environmental knowledge, attitudes, and behavior: A review and analysis of the literature and potential future research. *Journal of Sustainable Tourism*, *23*(6), 838–858. https://doi.org/10.1080/09669582.2015.1024258

Babintseva, E., Mansur, D., Trifonova, E., & Dugina, T. (2025). Impact of environmental education development on promoting environmentally safe use of natural resources and careful attitude toward nature. *International Journal of Ecosystems; Ecology Sciences, 15*(1), 123–132. https://doi.org/10.31407/ijees15.114

Ceballos, G., Ehrlich, P. R., Barnosky, A. D., García, A., Pringle, R. M., & Palmer, T. M. (2015). Accelerated modern human–induced species losses: Entering the sixth mass extinction. *Science Advances, 1*(5), e1400253. https://doi.org/10.1126/sciadv.1400253

Change, I. P. O. C. (2023). *Climate Change 2021—The Physical Science Basis: Working Group I Contribution to the Sixth Assessment Report of the Intergovernmental Panel on Climate Change*. Cambridge University Press. https://doi.org/10.1017/9781009157896

Crutzen, P. J. (2002). Geology of mankind. *Nature, 415*(6867), 23–23. https://doi.org/10.1038/415023a

Cruz-Quijada, L. G., & Tejedor-Flores, N. (2024, October 23–25). Experiences and actions on environmental education in solid waste management in Panama. 2024 9th International Engineering, Sciences and Technology Conference (IESTEC), Panama City, Panama.

Diaz, R. J., & Rosenberg, R. (2008). Spreading dead zones and consequences for marine ecosystems. *Science, 321*(5891), 926–929. https://doi.org/10.1126/science.1156401

Diffenbaugh, N. S., & Burke, M. (2019). Global warming has increased global economic inequality. *Proceedings of the National Academy of Sciences, 116*(20), 9808–9813. https://doi.org/10.1073/pnas.1816020116

Ding, H., Qiu, W., Deng, D., & Yan, M. (2025). Addressing challenges in on-campus and off-campus collaborative environmental education: Possible solutions from frontline teacher interviews. *Integrative Conservation, 4*(1), 33–37. https://doi.org/10.1002/inc3.70006

Doney, S. C., Fabry, V. J., Feely, R. A., & Kleypas, J. A. (2009). Ocean acidification: The other CO2 problem. *Annual Review of Marine Science, 1*, 169–192. https://doi.org/10.1146/annurev.marine.010908.163834

Dos Santos Castro, N., & Soares, A. L. M. (2024). Green market: A tool for recycling and environmental education tool. *Revista de Gestão Social e Ambiental, 18*(3), 1–15. https://doi.org/10.24857/rgsa.v18n3-060

Driscoll, C. T., Mason, R. P., Chan, H. M., Jacob, D. J., & Pirrone, N. (2013). Mercury as a global pollutant: Sources, pathways, and effects. *Environmental Science & Technology, 47*(10), 4967–4983. https://doi.org/10.1021/es305071v

Ferronato, N., & Torretta, V. (2019). Waste mismanagement in developing countries: A review of global issues. *International Journal of Environmental Research and Public Health, 16*(6), 1060. https://doi.org/10.3390/ijerph16061060

Hemby, T. L., Powell, R. B., & Stern, M. J. (2024). Availability and distribution of environmental education field trip programs for adolescent students in the US: A national study of spatial accessibility. *Environmental Education Research, 30*(2), 214–234. https://doi.org/10.1080/13504622.2023.2237706

Hildebrand, J. A. (2009). Anthropogenic and natural sources of ambient noise in the ocean. *Marine Ecology Progress Series, 395*, 5–20. https://doi.org/10.3354/meps

Jamieson, A. J., Brooks, L. S. R., Reid, W. D. K., Piertney, S. B., Narayanaswamy, B. E., & Linley, T. D. (2019). Microplastics and synthetic particles ingested by deep-sea amphipods in six of the deepest marine ecosystems on Earth. *Royal Society Open Science, 6*(2). https://doi.org/10.1098/rsos.180667

Jones, K. C., & de Voogt, P. (1999). Persistent organic pollutants (POPs): State of the science. *Environmental Pollution, 100*(1), 209–221. https://doi.org/10.1016/S0269-7491(99)00098-6

Karagulian, F., Belis, C. A., Dora, C. F. C., Prüss-Ustün, A. M., Bonjour, S., Adair-Rohani, H., & Amann, M. (2015). Contributions to cities' ambient particulate matter (PM):

A systematic review of local source contributions at global level. *Atmospheric Environment*, *120*, 475–483. https://doi.org/10.1016/j.atmosenv.2015.08.087

Karatayev, A. Y., Burlakova, L. E., & Padilla, D. K. (2015). Zebra versus quagga mussels: A review of their spread, population dynamics, and ecosystem impacts. *Hydrobiologia*, *746*(1), 97–112. https://doi.org/10.1007/s10750-014-1901-x

Lenton, T. M., Held, H., Kriegler, E., Hall, J. W., Lucht, W., Rahmstorf, S., & Schellnhuber, H. J. (2008). Tipping elements in the Earth's climate system. *Proceedings of the National Academy of Sciences*, *105*(6), 1786–1793. https://doi.org/10.1073/pnas.0705414105

Liu, J., Hull, V., Batistella, M., DeFries, R., Dietz, T., Fu, F., Hertel, T. W., Izaurralde, R. C., Lambin, E. F., Li, S., Martinelli, L. A., McConnell, W. J., Moran, E. F., Naylor, R., Ouyang, Z., Polenske, K. R., Reenberg, A., de Miranda Rocha, G., Simmons, C. S., Verburg, P. H., Vitousek, P. M., Zhang, F., & Zhu, C. (2013). Framing sustainability in a telecoupled world. *Ecology and Society*, *18*(2), Article 26. https://doi.org/10.5751/ES-05873-180226

Lozano-Lucia, O. R., & Talavera-Ortega, M. (2024). Enfoque de la educación ambiental en educación infantil: Estudio comparado en docentes de España y Costa Rica. *Revista Electrónica Educare*, *28*(2), 22–40. https://doi.org/10.15359/ree.28-2.18430

Münzel, T., Gori, T., Babisch, W., & Basner, M. (2014). Cardiovascular effects of environmental noise exposure. *European Heart Journal*, *35*(13), 829–836. https://doi.org/10.1093/eurheartj/ehu030

Persano Adorno, D., Birsan, E. A., Stoica, S. F., Capatina, M., Cojocaru, C., Tzortzaki, A., Štanfelj, Z., Dinçer, Y. S., & Pizzolato, N. (2025). Sustainability as a cross-curricular link: Creative European strategies for eco-conscious environmental education. *Sustainability*, *17*(11), 5193. https://doi.org/10.3390/su17115193

Pimm, S. L., Jenkins, C. N., Abell, R., Brooks, T. M., Gittleman, J. L., Joppa, L. N., Raven, P. H., Roberts, C. M., & Sexton, J. O. (2014). The biodiversity of species and their rates of extinction, distribution, and protection. *Science*, *344*(6187), 1246752. https://doi.org/10.1126/science.1246752

Pope III, C. A., Burnett, R. T., Thun, M. J., Calle, E. E., Krewski, D., Ito, K., & Thurston, G. D. (2002). Lung cancer, cardiopulmonary mortality, and long-term exposure to fine particulate air pollution. *JAMA*, *287*(9), 1132–1141. https://doi.org/10.1001/jama.287.9.1132

Rabalais, N. N., Turner, R. E., & Wiseman, W. J. (2002). Gulf of Mexico hypoxia, A.K.A. "The dead zone". *Annual Review of Ecology, Evolution, and Systematics*, *33*, 235–263. https://doi.org/10.1146/annurev.ecolsys.33.010802.150513

Ripple, W. J., & Beschta, R. L. (2012). Trophic cascades in Yellowstone: The first 15years after wolf reintroduction. *Biological Conservation*, *145*(1), 205–213. https://doi.org/10.1016/j.biocon.2011.11.005

Rockström, J., Steffen, W., Noone, K., Persson, Å., Chapin, F. S., Lambin, E. F., Lenton, T. M., Scheffer, M., Folke, C., Schellnhuber, H. J., Nykvist, B., de Wit, C. A., Hughes, T., van der Leeuw, S., Rodhe, H., Sörlin, S., Snyder, P. K., Costanza, R., Svedin, U., Falkenmark, M., Karlberg, L., Corell, R. W., Fabry, V. J., Hansen, J., Walker, B., Liverman, D., Richardson, K., Crutzen, P., & Foley, J. A. (2009). A safe operating space for humanity. *Nature*, *461*(7263), 472–475. https://doi.org/10.1038/461472a

Safardokkht bahar, P., Shobeiri, S. M., Rezaei, M., & Sarmadi, M. R. (2025). Application of meta-synthesis technique in environmental education policy based on carbon footprint management in gas refineries. *Journal of Environmental Studies*, *50*(4), 457–475. https://doi.org/10.22059/jes.2025.375867.1008498

Schuur, E. A. G., McGuire, A. D., Schädel, C., Grosse, G., Harden, J. W., Hayes, D. J., Hugelius, G., Koven, C. D., Kuhry, P., Lawrence, D. M., Natali, S. M., Olefeldt, D., Romanovsky, V. E., Schaefer, K., Turetsky, M. R., Treat, C. C., & Vonk, J. E. (2015).

Climate change and the permafrost carbon feedback. *Nature*, *520*(7546), 171–179. https://doi.org/10.1038/nature14338

Slabbekoorn, H., & Peet, M. (2003). Birds sing at a higher pitch in urban noise. *Nature*, *424*(6946), 267–267. https://doi.org/10.1038/424267a

Thompson, R. C., Olsen, Y., Mitchell, R. P., Davis, A., Rowland, S. J., John, A. W. G., McGonigle, D., & Russell, A. E. (2004). Lost at sea: Where is all the plastic? *Science*, *304*(5672), 838–838. https://doi.org/10.1126/science.1094559

Turetsky, M. R., Abbott, B. W., Jones, M. C., Anthony, K. W., Olefeldt, D., Schuur, E. A. G., Grosse, G., Kuhry, P., Hugelius, G., Koven, C., Lawrence, D. M., Gibson, C., Sannel, A. B. K., & McGuire, A. D. (2020). Carbon release through abrupt permafrost thaw. *Nature Geoscience*, *13*(2), 138–143. https://doi.org/10.1038/s41561-019-0526-0

2 Fundamentals of Environmental Education

2.1 DEFINITION AND HISTORICAL EVOLUTION OF ENVIRONMENTAL EDUCATION

2.1.1 Contemporary Conceptualization of Environmental Education

Environmental education is a complex, multidimensional field that has evolved from early concepts focused on nature conservation to sophisticated approaches that integrate ecological, social, economic, political, and ethical dimensions of sustainable development (Stevenson et al., 2013). In its contemporary conceptualization, environmental education transcends the simple transmission of knowledge about the environment to become a transformative process that develops citizens capable of understanding, evaluating, and acting on the complex interactions between social and natural systems.

Contreras Ameduri (2025) on "avian pedagogies: women ornithologists and environmental education in late 19th-century America" provides important historical perspectives on the origins of environmental education. Her analysis of ornithological children's literature by 19th-century female naturalists reveals how these works "paved the way for early forms of environmental education" and encouraged children's participation in early animal welfare and nature conservation movements, thus fostering the development of environmental awareness from an early age at a time of recurrent bird slaughter.

The contemporary definition of environmental education recognizes that environmental problems are not merely technical but fundamentally social, rooted in value systems, economic structures, and power relations that require transformation. This understanding has led to educational approaches that emphasize critical thinking, democratic participation, and collective action as essential components of environmental education.

Environmental education is distinguished from other educational fields by its emphasis on direct experience with the natural world, interdisciplinary learning that integrates multiple forms of knowledge, and an explicit commitment to social transformation toward more sustainable relationships between human societies and natural systems. These distinctive characteristics require innovative pedagogical methods to facilitate cognitive, affective, experiential, and action-oriented learning.

2.1.2 Historical Background and Formative Influences

The antecedents of environmental education can be traced to multiple intellectual traditions and social movements that converged in the 20th century, creating

DOI: 10.1201/9781003779667-3

conditions conducive to its emergence as a distinct discipline. The progressive education movement, led by John Dewey and other educational reformers, emphasized experiential learning and the connection between education and social reform, laying pedagogical foundations that later influenced approaches to environmental education (Gruenewald, 2003).

The conservation movement of the 19th and early 20th centuries, exemplified by figures such as John Muir and Aldo Leopold, articulated environmental philosophies that recognized the intrinsic value of nature and the need for more harmonious relationships between humans and the natural world. These philosophical perspectives provide important conceptual foundations for this study (Callicott, 1990).

Hesselink (2024), in "Caring for the Earth: Case Studies in Environmental Education and Communication," documents the evolution of integrated approaches that combine "what" and "how" of sustainable development. His analysis emphasizes that "integrating what and how sustainable development is the essence of environmental education and communication" and that "is a double-edged challenge" because "despite all our knowledge, it is unclear what equity, and sustainable living will mean for all the inhabitants of Earth a generation from now."

Outdoor education and youth camp movements have provided practical methods for facilitating direct experiences with nature, which have become central to many environmental education programs. These experiential approaches demonstrate the educational potential of immersive experiences in natural settings for developing appreciation, understanding, and commitment to conservation.

2.1.3 The Stockholm Conference and International Recognition

The United Nations Conference on the Human Environment, held in Stockholm in 1972, marked a milestone in the history of environmental education as an internationally recognized field (McKeown & Hopkins, 2003). The Stockholm Declaration established environmental education as essential for disadvantaged communities. This official declaration granted international legitimacy to environmental education and established fundamental principles that continue to influence contemporary policies.

2.1.4 The Belgrade Seminar and the Fundamental Objectives

The International Seminar on Environmental Education, held in Belgrade in 1975, articulated the fundamental objectives of environmental education, which remain central to this field (Tilbury, 1995). The Belgrade objectives include raising awareness of environmental problems, developing knowledge of the environment and associated problems, fostering concern for the environment, developing skills to solve environmental problems, providing opportunities for active participation in solving environmental problems, and developing the capacity to evaluate environmental education measures and programs.

These objectives reflect a holistic conceptualization of environmental education that integrates the cognitive, affective, and behavioral dimensions of learning. The emphasis on active participation and problem-solving distinguishes environmental education from traditional educational approaches that primarily focus on knowledge transmission.

2.1.5 The Tbilisi Conference and Conceptual Consolidation

The Intergovernmental Conference on Environmental Education, held in 1977, is often considered crucial because it established the first clear, organized, global definition of environmental education, along with its purpose and implementation in diverse educational and social settings. This conference produced guidelines that have been frequently referenced; they not only outlined general objectives but also translated those objectives into clear pedagogical criteria (perspectives, themes, levels of analysis, and ways to participate in the conference). The result was a framework broad enough to be applicable across countries, yet concrete enough to guide curricula, teacher training programs, and community projects.

In simple terms, the so-called "Tbilisi Principles" remain relevant, as they combine three key strengths: A holistic vision, a path to action, and social significance. First, they emphasize that the environment cannot be understood solely from a natural perspective but rather as a system encompassing ecological, economic, social, cultural, and political aspects of the environment. On the other hand, they state that environmental education must go beyond simply providing information; it must build the capacity to analyze problems, make decisions, participate, and act responsibly. Furthermore, they insist that learning must be integrated with daily life and the real-world challenges of each place, from the local to the global level (Marcinkowski, 2009).

It is helpful to detail the key points of these principles in this regard.

- Holistic and interdisciplinary approach: Incorporating natural sciences, social sciences, technology, and humanities to understand complex problems (e.g., water, waste, energy, and biodiversity) without breaking them down into isolated topics.
- Lifelong learning: Environmental education is not limited to the school phase; it is a continuous process that integrates formal, nonformal, and informal education.
- Contextual relevance: Topics must arise from the realities of the surrounding environment (problems, culture, history, and local resources) and be connected to regional and global perspectives.
- Critical thinking and problem-solving: Fostering diagnostic skills (identifying causes and consequences), evaluating options, and making informed decisions.
- Participation and shared responsibility: Encourage students and the community to be active participants, not passive observers; investigate, discuss, collaborate, and propose improvements to the project.
- Ethical and values-based orientation: Strengthen attitudes of care, intergenerational justice, respect for life, and social responsibility, avoiding a purely "technical" approach.
- Active learning: Preference is given to methods such as projects, case studies, fieldwork, school campaigns, environmental audits, and service learning.

That these principles "continue to provide essential guidance" means, in practice, that they still serve as good benchmarks for judging whether an environmental education plan is well-designed. For example, if a course only presents facts and concepts, Tbilisi encourages the inclusion of in-depth analysis and action; if a plan is very "ecological" but neglects the social aspect, the framework reminds us that human elements must be integrated; if the tasks lack context, it suggests looking at the specific location and the real problems.

They also explain why environmental education remains relevant in the face of current challenges (climate change, biodiversity loss, pollution, energy transition): Although the problems are more serious and the methods of teaching have changed (digital resources, blended learning, citizen science), the basic idea remains valid: Understanding complex systems, developing skills, and encouraging informed participation. In this sense, the 1977 principles are not just "an old piece of paper"; they remain a guide for creating educational materials that integrate knowledge, values, and action.

2.1.6 Evolution toward Education for Sustainable Development

The conceptualization of sustainable development, popularized by the 1987 Brundtland Report and institutionalized through the 1992 Rio Conference, significantly influenced the evolution of environmental education toward broader approaches that integrate environmental, social, and economic considerations into curricula (Hopwood et al., 2005). This evolution led to the development of education for sustainable development (ESD) as a framework that seeks to address the complexities of contemporary sustainability challenges in education.

Farliana and Hardianto (2024), in their research on evaluating the transition from environmental education to ESD from a global perspective in higher education, examine this conceptual evolution. Their analysis, based on a meta-analysis of 7,439 documents, found that ESD has been increasingly accepted globally since the decade was promoted by UNESCO.

The tension between traditional environmental education and sustainable development education continues to be debated in the literature. Some scholars argue that ESD dilutes the environmental focus, shifting toward a more comprehensive approach that addresses the systemic causes of environmental problems and requires social and economic transformation.

2.1.7 Influences of Critical and Liberating Pedagogies

Environmental education has been significantly influenced by critical and liberatory pedagogies, particularly Freire's work on education as a practice of freedom and the pedagogy of the oppressed. These approaches emphasize the importance of developing critical awareness of the social and economic structures that perpetuate both social injustice and environmental degradation (Misiaszek, 2015).

Critical pedagogies emphasize power analysis, democratic participation, and social transformation in the context of environmental education. These approaches recognize that environmental problems are often rooted in social and economic inequalities that must be addressed as an integral part of environmental solutions.

2.1.8 Incorporation of Indigenous and Traditional Knowledge

An important evolution in environmental education is the growing recognition of indigenous and traditional knowledge as valuable sources of environmental understanding and educational methodology. This recognition represents a shift away from predominant Western/modern approaches toward pluralistic perspectives that value various forms of knowledge. Zidny et al. (2020) conducted a comprehensive review of the literature on the integration of indigenous knowledge into science and environmental education. Their analysis found that "despite increasing recognition of indigenous knowledge in policy documents and curricula, significant barriers remain, including epistemological tensions between Western scientific and indigenous knowledge systems, insufficient teacher training, and institutional resistance to pluralistic approaches." The study concludes that environmental education can greatly benefit from incorporating indigenous knowledge, particularly in developing culturally relevant pedagogies that connect local ecological knowledge with scientific understanding, but that this integration requires systemic changes in curriculum design and teacher preparation.

2.1.9 Contemporary Developments and Future Directions

Contemporary developments in environmental education reflect responses to emerging environmental crises and the evolution of educational theory and practice. Climate change has emerged as a central concern, motivating the development of climate education as a specialized subfield within broader environmental education (Monroe et al., 2019).

The integration of digital technologies into environmental education represents one of the most significant contemporary developments in the field. Ardoin et al. (2013) conducted a systematic review of environmental education outcomes and their relationship with instructional design, including the role of technology-enhanced learning environments. More recently, Lowan-Trudeau (2023) examined the transformative potential of digital tools—including geographic information systems (GIS), citizen-science platforms, remote sensing data, and immersive virtual reality—for environmental education, concluding that "digital technologies offer remarkable possibilities for extending the scope and depth of environmental learning, but their effectiveness depends critically on pedagogical design that maintains authentic connections to local environments and avoids substituting virtual experiences for direct engagement with nature." These findings underscore that digital tools should complement, rather than replace, the experiential and place-based traditions that remain central to environmental education.

2.2 GUIDING PRINCIPLES AND FUNDAMENTAL OBJECTIVES

2.2.1 Holistic and Integrative Principles

The fundamental principles of environmental education reflect a holistic understanding of the relationships between social and natural systems, which require integrative educational approaches capable of addressing the complexities inherent in contemporary environmental problems. These principles transcend traditional disciplinary divisions to embrace systemic perspectives that recognize the interconnections among the ecological, social, economic, political, and cultural dimensions of sustainability.

The principle of comprehensiveness states that environmental education must consider the environment as a whole, including natural, social, economic, political, technological, historical, and cultural aspects. This comprehensive perspective requires educational programs to avoid the fragmentation that characterizes traditional disciplinary approaches and develop the capacity to understand and navigate complex systems in which multiple factors interact in a nonlinear manner.

The application of holistic principles in environmental education is well illustrated by geopark-based and place-based educational programs that integrate geological, ecological, and cultural knowledge. Farsani et al. (2011) examined how geoparks serve as living laboratories for environmental education, demonstrating that "the integration of geological heritage with biodiversity, cultural history, and sustainable management provides an ideal framework for holistic environmental learning." Their analysis of UNESCO Global Geoparks showed that educational programs embedded in these sites develop participants' ability to understand complex interactions among geomorphological processes, plant community composition, and human land-use practices, thereby supporting both environmental education and sustainable management objectives. These findings reinforce the principle that environmental education must consider the environment as a whole, connecting natural processes with social and economic dimensions.

2.2.2 Principle of Continuity and Ongoing Process

Environmental education is conceptualized as a continuous process that extends throughout life, recognizing that environmental understanding and the capacity for sustainable action require constant development and refinement in response to changes in scientific knowledge, environmental conditions, and social contexts. This principle of continuity distinguishes environmental education from one-off educational interventions and emphasizes the importance of creating adaptive and evolving learning systems. Olsson et al. (2022) provide compelling longitudinal evidence supporting this principle. Their multiyear study tracking secondary school students' sustainability consciousness demonstrated that environmental understanding develops progressively over time through sustained educational engagement and that the effects of environmental education are cumulative rather than instantaneous. Their findings confirmed that students who received continuous, curriculum-integrated sustainability education showed significantly stronger development across cognitive, affective, and behavioral dimensions compared to those receiving isolated interventions, underscoring that environmental education must be designed as an ongoing process rather than a series of discrete events.

The procedural nature of environmental education requires educational methodologies that facilitate self-directed learning, critical reflection, and continuous adaptation. The most effective programs are those that develop participants' ability to continue learning and independently evolve their environmental understanding, independent of specific formal education interventions.

2.2.3 Principle of Interdisciplinarity

The interdisciplinary nature of environmental problems requires educational approaches that integrate knowledge, methodologies, and perspectives from multiple disciplines in a coherent and pedagogically effective manner. Interdisciplinarity in environmental education goes beyond simply juxtaposing content from different disciplines to create genuine syntheses that address the complexity of socioecological systems.

Effective interdisciplinary implementation requires the development of competencies to navigate different disciplinary paradigms, communicate across fields, and synthesize information from diverse sources. Innovative approaches to interdisciplinary integration in environmental education have been increasingly supported by international collaborative projects. Gough (2002) provided an early and influential analysis of how environmental education can serve as a cross-curricular link, arguing that "environmental education provides a unique opportunity to integrate knowledge across disciplinary boundaries because environmental problems inherently require multi-perspective analysis." More recently, European collaborative projects, such as those funded under the Erasmus+framework, have sought to develop thematic environmental curriculum content that integrates across STEAM disciplines—science, technology, engineering, arts, and mathematics—enhancing multidisciplinary learning by connecting scientific knowledge with creative environmental practices. Hajer et al. (2015) emphasized that equipping students with the skills and mindset necessary for sustainable problem-solving requires moving beyond traditional disciplinary silos toward integrated curricula that reflect the systemic nature of environmental challenges.

2.2.4 Principle of Local–Global Contextualization

Environmental problems manifest at both local and global scales, requiring educational approaches that connect immediate local experiences to broader global processes and solutions. This principle recognizes that an effective understanding of environmental problems requires the ability to navigate multiple spatial and temporal scales and connect individual actions with systemic consequences.

Effective contextualization requires educational programs to begin with environmental problems immediately relevant to participants and progressively expand their understanding to include connections to regional, national, and global processes. This progression from local to global can make complex environmental problems more accessible by developing the capacity for systems thinking. Ardoin et al. (2020), in their systematic review of environmental education outcomes for conservation, found that programs grounded in local environmental contexts consistently produced stronger cognitive, affective, and behavioral outcomes than those relying on abstract or decontextualized content. Their analysis demonstrated that effective contextualization begins with participants' immediate environmental realities and progressively builds connections to regional and global processes, creating a scaffold that makes complex environmental challenges personally meaningful and actionable. These findings confirm that the progression from local to global scales is essential for developing systems thinking capacity in environmental education.

2.2.5 Objectives of Environmental Awareness Development

The development of environmental awareness is a fundamental goal of environmental education, encompassing the cognitive, affective, and ethical dimensions of understanding human–environment relationships. Environmental awareness transcends simple factual knowledge of environmental problems to encompass a deep understanding of systemic interconnections, sensitivity to nature's intrinsic values, and ethical responsibilities toward future generations and nonhuman species (Kollmuss & Agyeman, 2002).

For example, Leontev (2024), in the article "Environmental Education and Its Role in Shaping Environmental Knowledge and Attitudes," provides empirical evidence of these processes. Their study revealed that "environmental knowledge and students' attitudes toward the environment are positively correlated, but to a weak degree," suggesting that developing environmental awareness requires more than the simple transmission of information and must include approaches that can facilitate the transformation of attitudes and values.

2.2.6 Competency Development Objectives for Action

Environmental education seeks to develop practical competencies that enable participants to take effective action to address environmental problems in specific contexts. These competencies include technical skills for implementing environmental solutions, social skills for organizing collective action, and political skills for participating in decision-making processes that affect environmental quality (Jensen & Schnack, 1997).

Developing competencies for action requires opportunities for authentic practice, in which participants can experiment with different strategies, reflect on the results, and refine their approaches based on their experiences. The most effective programs provide platforms that allow participants to experiment with different strategies for environmental action, reflect on outcomes, and refine their approaches through iterative cycles of practice and reflection.

2.2.7 Objectives of Democratic Participation

Environmental education seeks to strengthen the capacity for democratic participation in environmental decision-making, recognizing that sustainable solutions require active environmental governance, advocacy, community organization, and engagement in formal political processes. Rousell and Cutter-Mackenzie-Knowles (2020), in their systematic review of climate change education, found that the most effective programs are those that give children and young people a genuine "voice" and a "hand" in redressing environmental challenges. Their analysis demonstrated that authentic democratic participation in environmental education requires moving beyond simulated exercises to create real opportunities for learners to engage in environmental advocacy, community organizing, and participation in decision-making processes that affect environmental quality in their communities.

Effective democratic participation requires not only technical knowledge of environmental issues but also an understanding of political processes, communication skills, and the ability to collaborate with citizens. The most effective environmental education programs provide opportunities for authentic practice in democratic participation through projects that involve real decision-making.

2.2.8 Social Transformation Objectives

A fundamental objective of environmental education is to contribute to social transformation toward more sustainable and equitable relationships between human societies and natural systems. This objective recognizes that environmental problems are often rooted in social, economic, and political structures that require systemic change rather than just modifications in individual behavior. Wals (2011) offers important perspectives on the transformative gap between acquiring pro-environmental knowledge and adopting pro-environmental behavior, arguing that "learning our way to sustainability requires a form of social learning that is transgressive, meaning it challenges and transcends existing norms, values, and practices rather than simply reinforcing them." He emphasizes the need to develop pedagogical and participatory alternatives in environmental education—including whole-school approaches, community-based action research, and deliberative dialogue—to address ecological dilemmas that cannot be resolved through information transmission alone.

2.2.9 Environmental Justice Objectives

Contemporary environmental education recognizes that environmental problems are unevenly distributed and disproportionately affect socially and economically marginalized communities (Mohai et al., 2009). Environmental justice objectives aim to develop an understanding of these inequalities and the capacity to work toward more equitable distributions of environmental benefits and burdens.

Developing awareness of environmental justice requires a critical analysis of how factors such as race, class, gender, and geography influence exposure to environmental problems and access to resources. It also requires building solidarity among affected communities and developing skills to advocate for environmental justice (Schlosberg, 2007).

2.2.10 Sustainable Development Goals

The Sustainable Development Goals integrate environmental concerns with social and economic development, recognizing that sustainability requires approaches that simultaneously address environmental protection, social equity, and economic development. This integration requires the ability to navigate complex trade-offs and develop solutions that optimize multiple objectives (Sachs, 2012).

Pereira et al. (2023) illustrate the importance of these integrated goals. In their research on perspectives on environmental education and justice for a post-COVID-19, Latin America and the Caribbean illustrate the importance of these integrated goals. Their study uses philosophical hermeneutics to identify perspectives that include adopting ecosocialism, changing the energy matrix, changing consumption practices, changing human–nature relationships, new ways of bringing nations together, overcoming the colonialist model, and strengthening potential and overcoming the discourse of poverty.

2.2.11 Ethical and Value Objectives

Environmental education aims to support the development and refinement of ethical frameworks to guide decision-making in contexts where environmental considerations must be balanced with other values and priorities. This includes exploring different environmental ethics, developing an appreciation for the intrinsic value of nature, and cultivating a sense of responsibility toward future generations.

The development of environmental ethics requires opportunities for deep reflection on values, engagement with diverse philosophical perspectives, and practice in applying ethical frameworks to specific environmental contexts. The most effective programs provide safe spaces for exploring values while challenging participants to expand their ethical considerations beyond their immediate self-interest.

2.3 THEORETICAL FRAMEWORK: FROM ENVIRONMENTAL AWARENESS TO BEHAVIORAL CHANGE

2.3.1 Theoretical Models of Behavioral Change

Understanding how environmental education can facilitate behavioral change toward more sustainable practices has evolved through the development of multiple theoretical frameworks that explain the complex relationships between knowledge, attitudes, intentions, and environmental behaviors. These theoretical models provide a conceptual foundation for designing educational interventions that maximize the likelihood of lasting behavioral changes.

The Theory of Planned Behavior, developed by Ajzen (1991), is one of the most influential frameworks for understanding the determinants of environmental behavior. This model proposes that behavioral intentions, which predict actual behavior, are determined by attitudes toward the behavior, subjective norms about social expectations, and perceived behavioral control, which reflects the ability to execute the behavior in question.

The Theory of Planned Behavior has been complemented by behavioral science insights that demonstrate how environmental education interventions can be enhanced through strategic combinations of cognitive and contextual strategies. Osbaldiston and Schott (2012) conducted a comprehensive meta-analysis of 253 experimental studies on pro-environmental behavior change and found that interventions combining cognitive strategies (such as providing knowledge and building commitment) with contextual modifications (such as prompts, social modeling, and goal-setting) produced significantly larger and more durable behavioral effects than either approach alone. Their analysis revealed that "interventions that engaged participants' intrinsic motivation while simultaneously providing structural support for behavior change were the most effective across diverse environmental domains, including energy conservation, recycling, and water use reduction." These findings suggest that environmental education programs for school-age children can maximize their impact by designing synergistic interventions that simultaneously boost environmental awareness and provide behavioral nudges that facilitate the translation of awareness into sustained action.

2.3.2 The Norm Activation Model

The Norm Activation Model, developed by Schwartz, focuses specifically on prosocial and altruistic behaviors, including many environmental behaviors that may require personal sacrifice for collective benefit (Schwartz, 1977). This model proposes that prosocial behavior is activated by awareness of the negative consequences for others, attribution of personal responsibility for these consequences, and the activation of personal norms that compel action.

This model is particularly relevant to environmental education because many environmental behaviors require individuals to act on behalf of beneficiaries who are distant in time and space, including future generations, nonhuman species, and communities in other geographical regions. Environmental education can help activate personal norms by raising awareness of environmental consequences, strengthening the sense of personal responsibility, and linking personal values to specific ethical obligations.

2.3.3 Dual-Processing Models

Dual-processing models recognize that human behavior is influenced by two cognitive processing systems: Automatic processing, which is fast, efficient, and often unconscious, and controlled processing, which is deliberative, conscious, and resource-intensive (Evans, 2008). These models have important implications for environmental education because they suggest that effective interventions must address both automatic responses and deliberative decision-making.

Automatic processing includes habits, emotional responses, and heuristics that can strongly influence environmental behavior, independent of formal knowledge or conscious intentions. Environmental education can modify automatic responses through immersive experiences that foster emotional connections with nature, develop environmental habits through repetitive practice, and modify environmental cues that trigger automatic behaviors. Whitburn et al. (2020), in their meta-analysis of human connection to nature and pro-environmental behavior, provide empirical support for this pathway. Their analysis found that emotional connection to nature—a construct rooted primarily in automatic, affective processing—is a significant and consistent predictor of pro-environmental behavior across diverse populations and environmental domains. These findings confirm that educational interventions targeting automatic emotional responses through direct nature experiences can be as effective as, or more effective than, interventions focused exclusively on deliberative knowledge acquisition.

2.3.4 Social Learning Theories

Social learning theories, particularly Bandura's social cognitive theory, emphasize the importance of observational learning, modeling, and social influence in shaping behavior (Bandura, 2001). These frameworks are particularly relevant to environmental education because environmental behavior is often socially embedded and shaped by norms, expectations, and the actions of others in social networks.

The concept of self-efficacy, which is central to social cognitive theory, is particularly important for environmental behavior because many environmental actions require confidence, observing role models who demonstrate environmental competence, and social support that reinforces the capability to act.

2.3.5 Stages of Change Models

Stages of change models, such as the Transtheoretical Model developed by Prochaska and DiClemente, propose that behavioral change occurs through predictable stages, each requiring different types of intervention (Prochaska & DiClemente, 1983). These models recognize that individuals may be at different stages of readiness for change and that effective interventions must be tailored to each participant's specific stage.

For environmental behavior, these stages typically include precontemplation (no awareness or intention to change), contemplation (awareness of problems but ambivalence about changing), preparation (intention to change and preliminary steps), action (implementation of environmental behaviors), and maintenance (sustained engagement in environmental practices). Educational interventions may be more effective when they align with participants' current readiness stages.

2.3.6 Theories of Diffusion of Innovations

Innovation diffusion theories explain how new ideas, technologies, and practices spread through social systems (Greenhalgh et al., 2004). These frameworks are relevant to environmental education because environmental behaviors often involve adopting innovative practices that may not be widely accepted or supported in the social contexts where individuals live and work.

Research on environmental innovations has identified the characteristics of innovations that predict their likelihood of adoption, including relative advantage over existing practices, compatibility with existing values and practices, simplicity of understanding and implementation, trialability on a small scale, and environmental observability.

2.3.7 Contextual and Structural Influences

While psychological models provide important insights into individual behavior change, research has increasingly recognized that environmental behavior is strongly influenced by contextual and structural factors that may enable or constrain individual choice. These factors include physical infrastructure, environmental incentives and barriers, social norms and support, and political context and cultural values (Stern, 2000).

Lozano et al. (2015) highlight the significance of contextual factors in contemporary environmental education programs within higher education. Their comprehensive review of declarations, charters, and partnerships related to sustainability in higher education found that while institutional commitments have increased substantially over the past two decades, "practical implementation remains uneven and is strongly mediated by institutional culture, leadership commitment, and the availability of resources for pedagogical innovation." Their analysis revealed that modern pedagogical approaches—including problem-based learning, interdisciplinary sustainability projects, campus-as-living-laboratory initiatives, and experiential community engagement—significantly enhanced students' engagement with environmental issues when embedded in supportive institutional structures that aligned curricular goals with operational practices.

2.3.8 Integration of Multiple Theoretical Perspectives

The most sophisticated frameworks for understanding environmental behavior change recognize that multiple theoretical perspectives offer complementary insights that can be integrated to yield a more comprehensive understanding of the subject. These integrated approaches acknowledge that environmental behavior is determined by complex interactions among psychological, social, and environmental factors that operate over multiple timescales.

Wamsler et al. (2021) provide a compelling framework for this integration. Their research on linking internal and external transformation for sustainability and climate action demonstrates that effective approaches must simultaneously address inner dimensions of change—including values, worldviews, emotional processing, and identity—alongside external dimensions such as institutional structures, social norms, and policy frameworks. Their proposed research and policy agenda emphasizes that neither purely psychological nor purely structural approaches are sufficient for achieving lasting environmental change in behavior and that the most promising interventions are those that create synergies between internal transformation and external enabling conditions.

Effective environmental education programs increasingly utilize multi-theoretical approaches that simultaneously address individual psychology, social dynamics, and structural constraints. These approaches recognize that sustainable behavior change requires interventions at multiple levels of the social-ecological system, from individual awareness and skills to policy change and social movement organization.

2.3.9 Implications for Educational Program Design

Insights from multiple theoretical frameworks have important implications for designing environmental education programs that aim to facilitate lasting behavioral change. Effective programs typically include environmental components that address different pathways to behavior change, including cognitive learning, affective development, skill building, social support, and structural modification (Hungerford & Volk, 1990).

The sequencing of these components is also important, with many effective programs starting with affective engagement that creates emotional connection and motivation, followed by cognitive learning that provides understanding and rationale, skill-building that develops practical competencies for environmental action, and social engagement that connects individual learning to collective action. This sequencing reflects insights from multiple theoretical frameworks suggesting that emotional engagement facilitates cognitive learning, which in turn supports the development of action competencies.

2.3.10 Measuring and Evaluating Behavioral Change

Measuring environmental behavior change presents unique challenges because many environmental behaviors are private, infrequent, or difficult to observe. Additionally, environmental attitudes and knowledge may change more readily than actual behaviors, making it difficult to evaluate the effectiveness of programs.

Effective evaluation of environmental education programs typically uses multiple measures of behavior change, including self-reports, behavioral observations, proxy measures such as resource consumption, and long-term follow-ups to assess the sustainability of changes. Mixed-methods approaches that combine quantitative measures with qualitative insights into participants' experiences and implementation barriers provide the most comprehensive understanding of program impacts.

2.4 ENVIRONMENTAL EDUCATION VERSUS EDUCATION FOR SUSTAINABILITY

2.4.1 Conceptual Evolution and Paradigmatic Tensions

The relationship between traditional environmental education and ESD has generated significant debate in environmental education, reflecting broader tensions regarding how to conceptualize the interplay between environmental protection, economic development, and social equity. These tensions are not merely academic; they have important practical implications for educational policy design, curriculum development, and resource allocation in environmental education programs.

Environmental education, as originally conceptualized at international conferences in the 1970s, focused primarily on developing awareness of environmental problems and the capacity for environmental protection and conservation. This approach reflects the historical context in which environmental degradation was perceived primarily as resulting from a lack of awareness of environmental processes and insufficient attention to environmental protection in decision-making (Wals, 2011).

Farliana and Hardianto (2024) provided evidence for this conceptual evolution. Their meta-analysis of 7,439 papers found that ESD's progress and EE's environmental context vary by geographic region, and that while ESD has become increasingly accepted globally since UNESCO's decade of promotion, EE has also played a more significant role in shaping the history of environmental education in certain regions.

2.4.2 Distinctive Approaches to Environmental Education

Traditional environmental education is characterized by several distinctive approaches that differentiate it from broader sustainability education approaches. First, environmental education typically emphasizes direct experience with natural environments, recognizing that emotional connections to nature are essential for developing a commitment to environmental protection.

Second, environmental education often emphasizes ecological literacy as the foundation for environmental understanding, developing comprehensive knowledge of ecological processes, biodiversity, and ecosystem functioning, which underpins the understanding of environmental problems and their solutions. This ecological focus distinguishes environmental education from approaches that prioritize social or economic dimensions of sustainability.

Third, environmental education typically maintains an explicit normative commitment to environmental protection and conservation, recognizing that education about the environment must also be education for the environment, seeking to develop environmental advocates and protectors. This advocacy orientation differs from ESD approaches that seek to maintain greater neutrality regarding the development paths.

2.4.3 Characteristics of ESD

ESD emerged from the recognition that environmental problems are often rooted in broader patterns of environmental development and social-environmental issues that require change beyond environmental protection alone. ESD seeks to integrate the environmental, social, and economic dimensions of sustainability into holistic educational approaches that address the root causes of unsustainability.

ESD typically emphasizes systems thinking and complexity more explicitly than traditional environmental education, recognizing that sustainable development requires an understanding of the complex interactions among multiple factors and trade-offs among different objectives. This systems approach can lead to a more sophisticated analysis of sustainability challenges but may also introduce greater complexity that is difficult to implement in education settings.

Castro Acevedo et al. (2023), in their paper entitled "Methodological Guidelines to Strengthen the Sociocultural Dimension in Community Environmental Education," illustrate approaches that bridge environmental education and sustainability education. Their methodology emphasizes "the importance of the approach taken" that "enriches traditional forms, enhances participation, critical awareness, and cooperation of the subjects who take on projects."

2.4.4 Debates on Scope and Priorities

One of the most significant debates between environmental education and ESD concerns the appropriate scope of educational interventions and the balance between environmental priorities and other developmental objectives. Traditional environmental education advocates argue that diluting the environmental focus with broader sustainability concerns may weaken the commitment to environmental protection and conservation (Jickling & Wals, 2008).

Proponents of ESD argue that environmental problems cannot be effectively addressed without addressing the social and economic factors that drive degradation. Education must prepare citizens to navigate the complex trade-offs between environmental, social, and economic objectives in real-world decision-making contexts. These debates reflect different assumptions about the relationship between environmental protection and economic development: Environmental education has traditionally emphasized the need to limit economic development that threatens environmental quality, while ESD seeks to identify development paths that can achieve environmental, social, and economic objectives simultaneously.

2.4.5 Distinctive Pedagogical Methodologies

Environmental education and ESD have developed somewhat different pedagogical approaches that reflect their different conceptual emphases on sustainability. Environmental education maintains a strong tradition of outdoor and experiential education, utilizing direct contact with natural environments as the primary vehicle for developing environmental understanding and commitment to it.

ESD is more likely to use simulations, case studies, and scenario planning to help participants understand the complex interactions among multiple sustainability factors by exploring different paths to sustainable development. These methodologies reflect ESD's emphasis on systems thinking and complexity management.

Chang and Chen (2023) highlight the practical implications of utilizing digital games for environmental education among school-age children, illustrating innovative pedagogical strategies that effectively integrate traditional environmental education with environment-enhanced learning. Their study found that "children have an improvement rate of about 60% in the implementation of environmental awareness knowledge and practice" through mobile game education.

2.4.6 Differences in Assessment and Outcomes

Environmental education and ESD also differ in their conceptualization and measurement of educational outcomes. Traditionally, environmental education has focused on outcomes such as environmental knowledge, ecological literacy, attitudes, and behaviors. These outcomes typically emphasize individual development in the environmental domain.

ESD typically uses a broader set of outcomes, including systems thinking capabilities, competencies for navigating complexity, and the ability to balance environmental objectives. ESD evaluation may also emphasize collective outcomes, such as community capacity building and institutional change, in addition to individual learning outcomes.

2.4.7 Complementarities and Opportunities for Synthesis

Despite the tensions between environmental education and ESD, many practitioners and researchers have identified important complementarities, suggesting opportunities for synthesis rather than competition between the approaches.

The integration of ESD frameworks into environmental education can help develop more sophisticated analyses of environmental problems while maintaining its distinctive strengths. Borba and Selles (2023) illustrate possibilities for synthesis in the study named "Natural History and Environmental Education in Dialogue in Science Teaching." Their analysis of pedagogical experiences that integrated natural history with environmental education demonstrates how different traditions can inform each other: "the experiences relate to a singular pedagogical action structured by building and maintaining a zoo where wild animals were protected and used in didactic activities under ecological-evolutionary perspectives and on behalf of the conservation of wild fauna."

2.4.8 Implications for Policy and Practice

The debates between environmental education and ESD have important implications for educational policy and practice, particularly in contexts where limited resources necessitate choices among the educational approaches. Policymakers must navigate between maintaining support for traditional environmental education and responding to the demand for broader sustainability education.

One approach adopted by many jurisdictions is to maintain environmental education as a distinct field while encouraging the integration of sustainability themes, where appropriate. This approach recognizes that different educational contexts and audiences may benefit from different emphases while avoiding the wholesale replacement of environmental education with ESD.

2.4.9 Future Directions and Emerging Opportunities

Looking forward, the relationship between environmental education and ESD will likely be characterized by increased dialogue and mutual learning rather than simple competition or replacement. Emerging environmental challenges, such as climate change, require both emotional engagement and ecological literacy, as emphasized by environmental education, as well as systems thinking and complexity management, as emphasized by ESD.

Future developments may include new synthetic approaches that combine the strengths of both methods and address their respective limitations (Wals & Benavot, 2017). These approaches may utilize environmental education. The study also suggests that context-specific approaches that tailor educational strategies to specific audiences, environments, and objectives may be the most effective. Rather than seeking universal solutions, practitioners may increasingly develop portfolios of approaches that include both environmental education and ESD strategies, depending on specific circumstances and objectives.

2.5 CONCLUSION

The analysis of the foundations of environmental education presented in this chapter reveals the conceptual richness and methodological complexity of an educational field that has evolved from initial conceptions focused on nature conservation to more sophisticated approaches that integrate multiple dimensions of sustainability. This evolution reflects both the development of scientific understanding of environmental problems and the refinement of educational theories and practices that can facilitate social transformation toward more sustainable relationships between human societies and natural systems.

The historical evolution of environmental education demonstrates that the field has been shaped by the convergence of multiple intellectual traditions and social movements, including progressive pedagogies, conservationist philosophies, critical pedagogies, and indigenous knowledge. This diversity of influences has created an educational field that is conceptually rich and complex, requiring careful navigation of different paradigms, methodologies, and objectives.

The guiding principles and fundamental objectives of environmental education established at international conferences in the 1970s continue to inform contemporary practice. However, they have been reinterpreted and expanded to reflect advances in the understanding of environmental problems and educational theory. The continuity of these fundamental principles suggests that they capture the essential aspects of environmental education that transcend specific historical contexts.

The theoretical framework for understanding the relationship between environmental awareness and behavioral change has evolved to recognize the complexity of psychological, social, and structural processes that influence environmental behavior. Insights from multiple theoretical perspectives suggest that effective environmental education must address the cognitive, affective, social, and structural dimensions of behavioral change, rather than relying solely on information transmission or awareness building.

The tension between traditional environmental education and ESD reflects broader debates about how to balance environmental protection, social equity, and development. While these tensions may seem problematic, they also provide valuable opportunities for dialogue and mutual learning that can strengthen both teaching approaches. Evidence suggests that synthesis approaches that combine the strengths of both traditions may be the most effective in addressing the complexity of contemporary sustainability challenges.

The holistic and integrative principles of environmental education require pedagogical methodologies that can facilitate interdisciplinary learning and develop environmental literacy to navigate complex systems in the future. This requirement presents both opportunities and challenges for educational practice, necessitating innovation in educational design while maintaining a focus on the fundamental objectives of environmental protection and social transformation.

The emphasis on local–global contextualization reflects the recognition that environmental problems manifest simultaneously on multiple scales, and effective environmental education must help participants understand the connection between local experiences and global processes. This multiscale approach requires sophisticated pedagogical capabilities but also provides opportunities to make environmental education more relevant and engaging for diverse audiences in the future.

The objectives of environmental development for action and democratic participation reflect an understanding that environmental education must go beyond environmental knowledge to develop the capabilities that enable effective environmental action and democratic participation. These action-oriented objectives distinguish environmental education from purely academic approaches and require methodologies that provide opportunities for authentic practice and reflection.

The integration of justice dimensions into environmental education objectives reflects a growing recognition that environmental problems are often rooted in social and economic inequalities that must be addressed alongside environmental solutions. This integration requires environmental education to develop competencies in social analysis and advocacy while maintaining a focus on environmental protection.

Theoretical frameworks for understanding behavioral change provide important guidance for designing educational interventions that can maximize the likelihood of generating sustainable changes in environmental behavior. However, these frameworks also reveal the complexity of behavioral change processes and the limitations of educational interventions that primarily operate at the individual level.

A comparison between environmental education and ESD illuminates both the opportunities and challenges of integrating environmental concerns with broader sustainability objectives. While tensions between these approaches are real, they also provide opportunities for dialogue and synthesis that can strengthen both fields.

Looking forward, the foundations of environmental education examined in this chapter suggest several important directions for future development in this field. First, the continued integration of diverse knowledge systems, including indigenous and traditional knowledge, can enrich environmental education while addressing issues of cultural relevance and social justice.

Second, innovation in pedagogical methods that facilitate interdisciplinary learning and systems thinking will become increasingly important as environmental problems become more complex and interconnected. This may require significant changes in educational institutions and teacher-training programs.

Third, strengthening the connection between environmental education and social movement organizations can help address the limitations of individual-focused approaches and contribute to the large-scale social transformation necessary to address environmental challenges.

Finally, continued research on the effectiveness of different educational approaches across contexts is essential for informing practice and policy while also contributing to the theoretical understanding of environmental learning processes. The foundations examined in this chapter provide a strong basis for understanding environmental education as both an educational practice and a social movement that seeks to transform human relationships with nature to promote sustainability. However, realizing this potential requires continued innovation, research, and commitment from diverse actors in educational and environmental communities.

REFERENCES

Ajzen, I. (1991). The theory of planned behavior. *Organizational Behavior and Human Decision Processes*, *50*(2), 179–211. https://doi.org/10.1016/0749-5978(91)90020-T

Ardoin, N. M., Bowers, A. W., & Gaillard, E. (2020). Environmental education outcomes for conservation: A systematic review. *Biological Conservation*, *241*, 108224. https://doi.org/10.1016/j.biocon.2019.108224

Ardoin, N. M., Clark, C., & Kelsey, E. (2013). An exploration of future trends in environmental education research. *Environmental Education Research*, *19*(4), 499–520. https://doi.org/10.1080/13504622.2012.709823

Bandura, A. (2001). Social cognitive theory: An agentic perspective. *Annual Review of Psychology*, *52*, 1–26. https://doi.org/10.1146/annurev.psych.52.1.1

Borba, R. C. D. N., & Selles, S. E. (2023). Natural history and environmental education in dialogue in science teaching. *Educação & Realidade*, *48*, e123937. https://doi.org/10.1590/2175-6236123937vs02

Callicott, J. B. (1990). Whither conservation ethics? *Conservation Biology*, *4*(1), 15–20. https://doi.org/10.1111/j.1523-1739.1990.tb00261.x

Castro Acevedo, G. D. P., Borges Machín, A. Y., Zurbano Cobas, L., & Mederos Jiménez, Y. (2023). Orientaciones metodológicas para fortalecer la Dimensión Sociocultural en la Educación Ambiental Comunitaria. Resultados: 2015–2020. *Universidad y Sociedad*, *15*(1), 251–258. https://rus.ucf.edu.cu/index.php/rus/article/view/3537

Chang, C.-L., & Chen, C.-N. (2023). Practical impact of school-age children using digital games to learn environmental education. E3S Web of Conferences,

Contreras Ameduri, C. (2025). Avian ecopedagogies: Women ornithologists and environmental education in late nineteenth-century America. https://doi.org/10.14198/raei.25702

Evans, J. S. B. T. (2008). Dual-processing accounts of reasoning, judgment, and social cognition. *Annual Review of Psychology*, *59*, 255–278. https://doi.org/10.1146/annurev.psych.59.103006.093629

Farliana, N., & Hardianto, H. (2024). Assessing the transition from environmental education to education for sustainable development: A global perspective in higher education. IOP Conference Series: Earth and Environmental Science.

Farsani, N. T., Coelho, C., & Costa, C. (2011). Geotourism and geoparks as novel strategies for socio-economic development in rural areas. *International Journal of Tourism Research*, *13*(1), 68–81. https://doi.org/10.1002/jtr.800

Gough, A. (2002). Mutualism: A different agenda for environmental and science education. *International Journal of Science Education*, *24*(11), 1201–1215. https://doi.org/10.1080/09500690210136611

Greenhalgh, T., Robert, G., Macfarlane, F., Bate, P., & Kyriakidou, O. (2004). Diffusion of innovations in service organizations: Systematic review and recommendations. *The Milbank Quarterly*, *82*(4), 581–629. https://doi.org/10.1111/j.0887-378X.2004.00325.x

Gruenewald, D. A. (2003). The best of both worlds: A critical pedagogy of place. *Educational Researcher*, *32*(4), 3–12. https://doi.org/10.3102/0013189x032004003

Hajer, M., Nilsson, M., Raworth, K., Bakker, P., Berkhout, F., De Boer, Y., Rockström, J., Ludwig, K., & Kok, M. (2015). Beyond cockpit-ism: Four insights to enhance the transformative potential of the sustainable development goals. *Sustainability*, *7*(2), 1651–1660. https://doi.org/10.3390/su7021651

Hesselink, F. (2024). Caring for the Earth: Case studies in environmental education and communication. In *A Sourcebook for Environmental Education: A Practical Review Based on the Belgrade Charter* (pp. 40–51). CRC Press. https://doi.org/10.1201/9781003579663

Hopwood, B., Mellor, M., & O'Brien, G. (2005). Sustainable development: Mapping different approaches. *Sustainable Development*, *13*(1), 38–52. https://doi.org/10.1002/sd.244

Hungerford, H. R., & Volk, T. L. (1990). Changing learner behavior through environmental education. *The Journal of Environmental Education*, *21*(3), 8–21. https://doi.org/10.1080/00958964.1990.10753743

Jensen, B. B., & Schnack, K. (1997). The action competence approach in environmental education. *Environmental Education Research*, *3*(2), 163–178. https://doi.org/10.1080/1350462970030205

Jickling, B., & Wals, A. E. J. (2008). Globalization and environmental education: Looking beyond sustainable development. *Journal of Curriculum Studies*, *40*(1), 1–21. https://doi.org/10.1080/00220270701684667

Kollmuss, A., & Agyeman, J. (2002). Mind the gap: Why do people act environmentally and what are the barriers to pro-environmental behavior? *Environmental Education Research*, *8*(3), 239–260. https://doi.org/10.1080/13504620220145401

Leontev, M. (2024). Environmental education and its role in the formation of environmental knowledge and attitudes. BIO Web of Conferences.

Lowan-Trudeau, G. (2023). Digital technologies and environmental education. *The Journal of Environmental Education*, *54*(1), 1–7. https://doi.org/10.1080/00958964.2022.2152413

Lozano, R., Ceulemans, K., Alonso-Almeida, M., Huisingh, D., Lozano, F. J., Waas, T., Lambrechts, W., Lukman, R., & Hugé, J. (2015). A review of commitment and implementation of sustainable development in higher education: Results from a worldwide survey. *Journal of Cleaner Production*, *108*, 1–18. https://doi.org/10.1016/j.jclepro.2014.09.048

Marcinkowski, T. J. (2009). Contemporary challenges and opportunities in environmental education: Where are we headed and what deserves our attention? *The Journal of Environmental Education*, *41*(1), 34–54. https://doi.org/10.1080/00958960903210015

McKeown, R., & Hopkins, C. (2003). EE p ESD: Defusing the worry. *Environmental Education Research*, *9*(1), 117–128. https://doi.org/10.1080/13504620303469

Misiaszek, G. W. (2015). Ecopedagogy and citizenship in the age of globalisation: Connections between environmental and global citizenship education to save the planet. *European Journal of Education*, *50*(3), 280–292. https://doi.org/10.1111/ejed.12138

Mohai, P., Pellow, D., & Roberts, J. T. (2009). Environmental justice. *Annual Review of Environment and Resources*, *34*, 405–430. https://doi.org/10.1146/annurev-environ-082508-094348

Monroe, M. C., Plate, R. R., Oxarart, A., Bowers, A., & Chaves, W. A. (2019). Identifying effective climate change education strategies: A systematic review of the research. *Environmental Education Research*, *25*(6), 791–812. https://doi.org/10.1080/13504622.2017.1360842

Olsson, D., Gericke, N., & Boeve-de Pauw, J. (2022). The effectiveness of education for sustainable development revisited: A longitudinal study on secondary students' action competence for sustainability. *Environmental Education Research*, *28*(3), 405–429. https://doi.org/10.1080/13504622.2022.2033170

Osbaldiston, R., & Schott, J. P. (2012). Environmental sustainability and behavioral science: Meta-analysis of proenvironmental behavior experiments. *Environment and Behavior*, *44*(2), 257–299. https://doi.org/10.1177/0013916511402673

Pereira, V. A., da Silva, R. F., & Valdes, A. D. J. N. (2023). Perspectives on environmental education and environmental justice for a Latin America and the Caribbean post-covid-19. *Universidad y Sociedad*, *15*(2), 73–81. https://rus.ucf.edu.cu/index.php/rus/article/view/3607

Prochaska, J. O., & DiClemente, C. C. (1983). Stages and processes of self-change of smoking: Toward an integrative model of change. *Journal of Consulting and Clinical Psychology*, *51*(3), 390–395. https://doi.org/10.1037/0022-006X.51.3.390

Rousell, D., & Cutter-Mackenzie-Knowles, A. (2020). A systematic review of climate change education: Giving children and young people a 'voice' and a 'hand' in redressing climate change. *Children's Geographies*, *18*(2), 191–208. https://doi.org/10.1080/14733285.2019.1614532

Sachs, J. D. (2012). From millennium development goals to sustainable development goals. *The Lancet*, *379*(9832), 2206–2211. https://doi.org/10.1016/S0140-6736(12)60685-0

Schlosberg, D. (2007). *Defining Environmental Justice: Theories, Movements, and Nature*. Oxford University Press. https://doi.org/10.1093/acprof:oso/9780199286294.001.0001

Schwartz, S. H. (1977). Normative influences on Altruism11 This work was supported by NSF Grant SOC 72–05417. I am indebted to L. Berkowitz, R. Dienstbier, H. Schuman, R. Simmons, and R. Tessler for their thoughtful comments on an early draft of this chapter. In L. Berkowitz (Ed.), *Advances in Experimental Social Psychology* (Vol. 10, pp. 221–279). Academic Press. https://doi.org/10.1016/S0065-2601(08)60358-5

Stern, P. C. (2000). New environmental theories: Toward a coherent theory of environmentally significant behavior. *Journal of Social Issues, 56*(3), 407–424. https://doi.org/10.1111/0022-4537.00175

Stevenson, R. B., Brody, M., Dillon, J., & Wals, A. E. J. (2013). *International Handbook of Research on Environmental Education* (1st ed.). Routledge. https://doi.org/10.4324/9780203813331

Tilbury, D. (1995). Environmental education for sustainability: Defining the new focus of environmental education in the 1990s. *Environmental Education Research, 1*(2), 195–212. https://doi.org/10.1080/1350462950010206

Wals, A. E. J. (2011). Learning our way to sustainability. *Journal of Education for Sustainable Development, 5*(2), 177–186. https://doi.org/10.1177/097340821100500208

Wals, A. E. J., & Benavot, A. (2017). Can we meet the sustainability challenges? The role of education and lifelong learning. *European Journal of Education, 52*(4), 404–413. https://doi.org/10.1111/ejed.12250

Wamsler, C., Osberg, G., Osika, W., Herndersson, H., & Mundaca, L. (2021). Linking internal and external transformation for sustainability and climate action: Towards a new research and policy agenda. *Global Environmental Change, 71*, 102373. https://doi.org/10.1016/j.gloenvcha.2021.102373

Whitburn, J., Linklater, W., & Abrahamse, W. (2020). Meta-analysis of human connection to nature and proenvironmental behavior. *Conservation Biology, 34*(1), 180–193. https://doi.org/10.1111/cobi.13381

Zidny, R., Sjöström, J., & Eilks, I. (2020). A multi-perspective reflection on how indigenous knowledge and related ideas can improve science education for sustainability. *Science & Education, 29*(1), 145–185. https://doi.org/10.1007/s11191-019-00100-x

3 Environmental Education as a Strategic Response

3.1 IMPACT OF EDUCATION ON SOCIAL TRANSFORMATION

3.1.1 Theoretical Foundations of Social Transformation through Education

The transformative dimension of environmental education is best understood through theoretical approaches that explain how it influences the broader social dynamics (Wamsler et al., 2021). In this vein, Foust et al. (2025) provide evidence of how certain curricular devices, such as alignment thinking and complementary curricula, can serve as levers for change amid increasing demand for environmental education. Based on a qualitative study of 13 teachers and administrators in an urban district, the authors explored how these strategies help to reduce implementation gaps in environmental education.

The results are presented in three sections. First, when there are gaps in teacher training and no explicit curriculum guidelines, teachers tend to respond with practical solutions: They adopt complementary materials and gradually incorporate an ecological perspective into their teaching. This suggests that environmental education can drive pedagogical innovation, even in contexts with limited institutional support.

Second, the prior training experiences recounted by the participants demonstrate the value of alignment thinking as a means to facilitate the immediate integration of environmental content in the classroom. These experiences appear to have long-term effects, persistently influencing the pedagogical decisions and professional development of teaching staff. Third, the analysis suggests that adopting an ecological equity perspective directly could strengthen both teaching and knowledge acquisition. From this perspective, ecological education not only seeks to promote ecological action but also connects with debates on community equality, providing tools to highlight and combat the fundamental inequalities that underlie current environmental problems (Menton et al., 2020).

Mishra et al. (2025) address sustainable environmental education by highlighting school-community collaboration. Through a case study at a secondary school in India, they analyzed how institutional leadership and coordination with community actors enabled the design and implementation of an environmental education plan focused on sustainability.

The data highlight the key role of educational leadership in strengthening internal dynamics, promoting social involvement, and strengthening ecological awareness among the different actors involved. In summary, the evidence suggests that paths to social change linked to environmental learning typically require leaders

 DOI: 10.1201/9781003779667-4

with a global perspective capable of logically linking the didactic, collective, and ecological.

In addition, the study highlights three enabling conditions: Visionary leadership, strengthened teaching capacity, and participatory governance. These elements, working in coordination, support an institutional transformation that extends the impact of environmental education beyond the classroom and promotes sustained change at the organizational and community levels.

3.1.2 Psychological Mechanisms of Behavioral Change

Analyzing the psychological mechanisms involved in behavioral change is key to understanding and expanding the transformative potential of environmental education programs. From this perspective, Xie and Yang (2025) examined the use of immersive virtual reality in environmental education settings, focusing on three processes: The development of environmental empathy, skill transfer, and attitude shaping.

The experiment was conducted with sixth-grade students and compared virtual reality experiences and traditional methods. The data showed increases in environmental empathy in both groups; however, those who participated in immersive virtual reality showed greater increases.

A particularly relevant finding is that, within the virtual reality group, environmental awareness and empathy had the greatest influence on the formation of environmental attitudes. This indicates that immersion can strengthen the mental links among cognitive understanding, emotional involvement, and the desire to act, thereby increasing the transition from learning to pro-environmental behavior.

Therefore, environmental empathy is a key intermediary between the learning experience and behavior. Rather than simply providing information, environmental education can increase its impact by fostering genuine emotional relationships with nature, which tend to sustain lasting behavioral changes (Whitburn et al., 2020).

Similarly, Chen and McCreary (2025) examined the impact of environmental education on attitudes toward nature in homeschooled students. Their exploratory study combined pre- and post-intervention quantitative measurements with qualitative evidence to detect changes in perceptions and attitudes after participating in an educational program.

Their findings point to a strengthening of the connection with nature and a favorable trend in environmental perceptions. However, statistically significant changes were limited, with quantitative measures showing little variation in eco-affinity and behavioral intentions. Taken together, these results highlight the complexity of psychological change and support the need to design environmental education programs that integrate cognitive, emotional, and behavioral components in a coordinated manner to maximize their transformative effects (Ardoin et al., 2020).

3.1.3 Transformation of Identities and Values

Environmental education has a distinctive capacity to shape identities and value frameworks that endure throughout one's life cycle. In this regard, Karmakar (2025), who focuses on transformative learning based on the figure of Wangari Maathai, analyzes the role of picture books as pedagogical resources for promoting environmental education and sustainability through narratives with high symbolic content.

The analysis of the picture book Seeds of Change: Planting a Path to Peace shows how this type of story effectively conveys Wangari Maathai's environmental and social activism (Johnson, 2010). According to the author, this text is an "informative picture book with an ecological focus" that catalyzes transformative learning by strengthening ecological literacy in childhood and simultaneously serves as an accessible means of disseminating principles of environmental education and sustainability to wider audiences.

The influence of narratives in shaping identity is particularly relevant in environmental education, as it enables students to conceive themselves as agents capable of addressing environmental problems. Identification with role models such as Wangari Maathai fosters self-images associated with social responsibility and the possibility of actively intervening to protect the environment.

In addition to these contributions, Slikkerveer and Gellaerts (2024) developed an ethno-scientific model of eco-integrated education based on indigenous knowledge systems. Their transdisciplinary proposal aims to reduce the so-called "knowledge gap" between academic science and ethno-science, with the goal of educating global citizens with solid ecological awareness.

This approach is based on the recognition that identity transformation involves articulating various forms of knowledge and value systems. When students can link scientific knowledge with traditional wisdom and diverse cultural perspectives, they develop more complex and reflective identities that can address the multifaceted nature of contemporary environmental challenges (Fernández-Llamazares et al., 2021).

3.1.4 Scaling Up from Individual to Collective Transformation

One of the central challenges of environmental education is ensuring that the processes of change generated at the individual level translate into sustained transformations in larger social systems (O'Brien, 2018). In this context, Ballard et al. (2024), who focused on the outcomes of environmental education grounded in community and citizen science, provide relevant evidence for understanding the mechanisms by which such processes can be expanded and consolidated.

Based on a systematic review of empirical studies, the authors found that community and citizen-science approaches to environmental education generally yield favorable results in both adult and youth populations. Specifically, significant progress was identified at various learning levels, including increased mastery of scientific content, strengthened research skills, consolidated positive attitudes toward science and the environment, and increased self-efficacy in understanding and addressing scientific and environmental issues.

Despite this, the observed effects extend beyond the individual level. The study also provided strong evidence of clear improvements in pro-environmental actions and in caring for and managing the environment, along with strengthened ties between neighbors and ways of working together in society. These data show that environmental education, when implemented through methods that involve everyone, not only helps each person learn but also encourages the development of skills among all and the formation of social networks that can sustain collaborative environmental projects in the future.

Furthermore, the findings show that community and citizen-science projects are most effective when they encourage participants to be fully involved in all stages of the research process. This includes not only data collection but also planning activities, analyzing and interpreting the information obtained, and disseminating the results of the study. In this sense, successful scaling-up processes seem to depend largely on creating spaces for genuine participation, in which social actors play a significant role in both knowledge production and decision-making related to environmental monitoring and management (Fraisl et al., 2022).

3.1.5 Institutional Factors That Facilitate Transformation

The processes of social transformation driven by environmental education do not develop in isolation; they depend largely on institutional environments that support and promote such changes. In this context, the study by Asli et al. (2024), which analyzes the effects of teacher participation in eco-pedagogy workshops, provides relevant empirical evidence of the role of institutional factors in promoting educational, environmental, and organizational approaches.

This study brought together several aspects and involved 44 primary school teachers. Participants in an eco-pedagogy program were compared with nonparticipants. The results showed that the participating teachers better understood and applied ideas related to environmental education. They also improved in the area of eco-pedagogical organization. Attending the workshops helped improve their understanding of this form of teaching. In addition, they improved their thinking and learned more about the subject.

A key finding was that attending specialized workshops is essential for teacher improvement. This shows that to truly change society, it is necessary to invest in strengthening institutions. This is important for environmental education to be implemented well and consistently.

In addition, the results showed that the teachers who participated in the program had a better understanding of the organizational ideas of eco-pedagogy. This means that effective environmental education not only changes how teachers teach but also shapes how students learn. It can also help schools change, become better at trying new things, and continue to change over time (Redman et al., 2018).

3.2 SCIENTIFIC EVIDENCE OF BEHAVIORAL CHANGE THROUGH EDUCATION

3.2.1 Methodologies for Measuring Behavioral Change

Producing solid scientific evidence of behavioral changes associated with environmental education requires advanced methodological approaches that account for the inherent complexity of these processes (Grimmette, 2014). In this regard, the study by Yin et al. (2025), which aimed to analyze the role of environmental education and the emotion of wonder in promoting pro-environmental behaviors among tourists, exemplifies methodological innovation in this field. This study analyzed 300 questionnaires and applied a structural equation modeling (SEM) with partial least squares (PLS) to compare two models in the Jinggangshan Scenic Area in China. While Model A focused on environmental education, Model B gave wonder a central role in triggering behavior change.

The results showed that both approaches significantly promoted pro-environmental behavior among visitors, particularly when environmental education components were combined with emotional engagement mechanisms. However, Model B, in which the experience of wonder predominates, has greater explanatory power. Specifically, the study showed that wonder and the so-called "need for accommodation" can directly stimulate pro-environmental behaviors among visitors and, through the mediating role of environmental education, indirectly stimulate them.

Likewise, the "evaluation of vastness" has a substantial influence on pro-environmental behavior, primarily through wonder and a sequential process that integrates emotional experiences with environmental learning. Taken together, these findings led the authors to propose a cognitive-behavioral framework that integrates environmental education and experiences of wonder to explain the mechanisms underlying behavioral change more accurately.

In a complementary approach to methodological innovation, Gilleran Stephens et al. (2025) analyzed universal design for learning (UDL) as a framework for redesigning environmental education outreach programs, offering valuable insights into how inclusive pedagogical design can enhance both the reach and effectiveness of environmental education initiatives. This research focused on identifying the main obstacles to the effective integration of environmental education in schools and analyzing the potential of Science Festivals designed using the principles of UDL to support teachers and students in overcoming these limitations.

A joint examination of survey and interview data revealed that environmental education is often relegated to the background in favor of other educational priorities, primarily due to curriculum overload, limited time, reductive conceptualizations, and logistical challenges.

However, following the implementation of the H_2O Science Festival, which was structured under the UDL approach, the results showed that experiential and practical learning proposals led to a significant increase in participation and involvement among teaching staff and students.

3.2.2 Experimental Evidence on Educational Effectiveness

Experimental research provides strong evidence that environmental education can produce observable behavioral changes. In this context, Hankin and Hogarth (2025) focused on the notion of becoming wild through the use of everyday materials such as chalk and paintbrushes, proposing a profound reconfiguration of environmental education pedagogies. From a posthumanist, feminist, and materialist perspective, the authors document how these objects cease to be mere teaching tools and become active participants in the production of knowledge by "thinking with" them rather than "about" them through processes of mutual interaction.

This pedagogical approach questions the anthropocentric, developmentalist, and civilizing logics that have historically characterized many educational models and proposes alternative forms of learning that allow "becoming wild" to serve as a pedagogical act. The findings suggest that this open and generative pedagogy recognizes the value of the everyday, the marginal, and the unconventional, expanding subjects' responsiveness and promoting more responsible relationships between human beings and the material worlds with which they coexist. In this sense, environmental education appears to be a space capable of redefining notions of agency and responsibility beyond the individual human subject.

Complementarily, Kraftl et al. (2025) explored environmental learning based on trees as the articulating axis of the educational process. Their transdisciplinary project, developed on a large scale with the participation of 545 children and young people from various primary and secondary schools in England and Scotland, analyzes how "starting with trees" can help us rethink the foundations of environmental education.

The results indicate that generating knowledge directly related to trees enables us to go beyond traditional approaches to "education for sustainability." The authors offer a substantive critique of dominant educational discourses, which tend to reproduce extractivist approaches and prioritize anthropocentric views that separate human experience from the rest of nature. These approaches, they argue, limit the understanding of the profound interdependence that exists between people and the ecological systems of which they are a part. In contrast to this approach, the project introduces the concepts of inhabiting, enabling, and belonging as an alternative framework for guiding educational processes that begin with the trees (Rousell & Cutter-Mackenzie-Knowles, 2020).

Taken together, these contributions show that environmental education can promote broader relationships between humans and nature, challenging the traditional assumption that focuses exclusively on humans. In doing so, they open up pedagogical horizons that recognize the interrelationships among people, other living organisms, and material environments, fostering a more integrated and relational understanding of contemporary environmental challenges.

3.2.3 Longitudinal Analysis of Persistence of Effects

In this regard, the work of Tyler (2025), focused on the contemporary environmental education imperative, provides a comprehensive examination of how the field has evolved in response to escalating ecological crises and the growing recognition that sustained educational efforts are essential for long-term environmental protection. Drawing on evidence from multiple international contexts, the author argues that the persistence of environmental education effects depends critically on the continuity and coherence of educational policies, the integration of environmental learning across all stages of formal and nonformal education, and the alignment of educational goals with broader societal commitments to sustainability.

The study indicates that environmental education greatly helps protect our planet. This also indicates that education is crucial for improving environmental management. By teaching people what they need to know and how to act, education helps address current problems and prepare for the future. Therefore, education not only informs but also gradually changes things over time.

Looking ahead, environmental education is essential for protecting the environment on an ongoing basis. This helps prevent new problems and solves existing problems. The goal is to achieve development that protects the environment. To determine whether this works, long-term research is required, following people for years or even decades after they learn about the subject. An illustrative example of this type of longitudinal analysis can be found in the study by Frankie and Vinson (2004) on conservation and environmental education in rural areas of northwestern Costa Rica. Based on the systematization of a nongovernmental organization's experiences, the authors identified a series of factors that explain the persistence of educational effects in rural communities.

The results show that the sustained effectiveness of environmental education programs depends largely on the degree of community ownership of initiatives, their alignment with local knowledge systems, and the strengthening of local institutional capacities. These elements allow educational actions to continue generating benefits even after external support ends, consolidating lasting impacts on community behavior and environmental management (Chawla, 2020).

3.2.4 Individual Differences in Educational Response

Research on behavioral change has shown that people's responses to environmental education programs are far from uniform, highlighting significant individual differences in learning, perception, and environmental awareness. Recognizing and analyzing these variations is key to designing tailored and effective educational strategies that respond to the diversity of educational backgrounds and contexts in which interventions occur.

In this regard, the study by Blasco-Magraner et al. (2025) focused on the use of soundscapes as a resource in environmental education, and this analysis examined the skills of teachers-in-training in auditory analysis and emotional responses to sound stimuli. The study was conducted with 151 teacher-training students: 62 in music education and 89 in general education (GE). The study examined how they noticed, described, and expressed emotional reactions to various environmental sounds to assess their ability to understand environmental sound information.

Overall, the data showed a limited ability to accurately detail the parts of the sound stimuli. Nevertheless, the group studying music education outperformed the GE group. In addition, a similar emotional pattern was observed: Sounds associated with natural places tended to elicit positive emotions, whereas those associated with urban environments tended to elicit negative emotions.

In this respect, no statistically significant differences were found between the two groups. Overall, the results support the usefulness of measuring sound awareness and sound environment analysis in university curricula, not only as an aid to strengthen environmental education but also as a potential means of enhancing emotional well-being. Simultaneously, evidence suggests that professional specialization significantly impacts how environmental information is understood and used, reinforcing the need to develop teaching styles that are context-sensitive and responsive to different learning styles.

3.2.5 Contextual Factors That Modulate Effectiveness

The available evidence on behavioral change indicates that the effectiveness of environmental education depends largely on contextual factors that can enhance or limit its impact. In this regard, the study by Tiemann and Ring (2025), which focuses on quantifying and mapping the provision of forest-related environmental education and training in Saxony, Germany, provides relevant information on the influence of these factors.

The study assessed the population's access to environmental education and training initiatives related to forest ecosystems, using a 90-minute maximum travel time as the reference, both by private and public transportation. The results indicate that when using individual transportation, all residents of Saxony can access this type of training within the established limit; however, when travel depends exclusively on public transportation, this possibility is reduced to less than half of the population.

Similarly, spatial analysis revealed significant territorial disparities in public transportation access. There was a marked concentration of educational and cultural resources in urban and peri-urban areas, particularly in the cities of Dresden and Leipzig. In contrast, other areas, such as Chemnitz, had significantly lower availability. In general, the evidence suggests that the availability of specific means and conditions for accessing ecological training processes is crucial, as it determines both the degree of involvement and knowledge construction.

Added to these considerations are the contributions of Rubáš et al. (2024), who examined the role of sense of place in geography and environmental education. Based on a review of empirical studies, the authors conclude that environmental- and place-based education approaches are effective in fostering a sense of place in students.

The analysis also points out that a clear conceptualization of sense of place, along with an empirical approach, facilitates its effective incorporation into geography and environmental education. This approach strengthens the link between students and their environment, a central element in developing geographic and environmental literacy and in forming attitudes and behaviors more committed to caring for the environment.

3.3 CONTRIBUTION TO THE SUSTAINABLE DEVELOPMENT GOALS (SDGS)

3.3.1 Integration of Environmental Education with the 2030 Agenda

Environmental education plays a strategic role in the 2030 Agenda for Sustainable Development, particularly through SDG 4 (Quality Education) and Target 4.7, which aims to ensure that students acquire the knowledge, skills, and values necessary to promote sustainable development (Kopnina, 2020). In this context, the study by Bradecki et al. (2024), regarding sustainable education through design, offers a concrete example of how environmental education can contribute directly and simultaneously to various SDGs.

This research examined ways to construct environmental education sites in an eco-friendly style. Therefore, the authors propose clear guidelines for assessing the eco-friendliness of buildings. They showed that these sites, with many different classes, serve as special places for developing comprehensive, eco-friendly teaching.

Incorporating sustainability ideas into educational buildings is a clear example of how environmental education can have effects beyond teaching, while also helping to create sustainable cities and communities (SDG 11), support responsible consumption and production patterns (SDG 12), and take action against climate change (SDG 13).

In addition, Purnomo et al. (2024), who focused on a prophetic form of environmental education and community power, provided further evidence of embedding sustainability in clear sociocultural settings. Through a study of how pesantren grow sustainably, the authors explain that the concept of "Green Pesantren" combines environmental education with Islamic lessons and social values, resulting in a balanced, appropriate approach to teaching.

The results show that implementing programs such as Adiwiyata and Ekopesantren strengthened environmental awareness and promoted sustainability within the pesantren community. This experience shows that environmental education can fit well with cultural and religious value systems, contributing not only to quality education (SDG 4) but also to poverty reduction (SDG 1) and sustainable cities and communities (SDG 11).

3.3.2 Specific Contribution to Educational SDGs

SDG 4 aims to ensure inclusive, equitable, and high-quality education and promote lifelong learning opportunities for all. Within this framework, environmental education makes relevant contributions that go beyond the isolated incorporation of ecological content into curricula, influencing broader pedagogical, social and cultural dimensions.

From this perspective, Massuchin et al. (2025) explored the relationship between environmental and health education using a phenomenological ecological approach and provided a theoretical analysis that demonstrates this multidimensional contribution. The authors examine the link between the two educational fields through a thought-provoking pedagogical vision. They indicate that the study should be understood as a complete whole, in which educational experiences are created by bringing people together with the places they live. As a place to come together, they offer the idea of "Living Well," which brings together ecological and health education. This way of looking at environmental and health education through an ecophenomenological lens highlights the potential of holistic approaches that transcend disciplinary boundaries, positioning educational experiences as embodied encounters between learners and the living environments they inhabit.

The integration of environmental and health education highlights the potential of comprehensive approaches to improve human well-being by simultaneously addressing the cognitive, emotional, social, and environmental aspects of learning (Tillmann et al., 2018). Additional research by Roy (2024) provides evidence of the contribution of environmental education to improving truly inclusive education, particularly in special education. Their research examines the strategic role of this field of education in raising environmental awareness, fostering positive attitudes toward nature, and introducing sustainable practices appropriate for different learning contexts.

The results highlight the need to develop inclusive environmental education services that meet the diverse needs of students, especially those with disabilities. Through sensory experiences in natural environments and hands-on teaching methods, special education fosters deeper and more meaningful connections between students and their environment, thereby reinforcing the principles of equity and quality education promoted by SDG 4.

3.3.3 Alignment with Environmental SDGs

Environmental education is directly related to several environmental SDGs, including access to clean water and sanitation (SDG 6), promotion of clean and accessible energy (SDG 7), action on climate change (SDG 13), conservation of marine ecosystems (SDG 14), and protection of terrestrial ecosystems (SDG 15) (Kopnina, 2020; Markkanen & Anger-Kraavi, 2019). These contributions are evident in studies that integrate educational processes with conservation practices and sustainable use of natural resources.

In this regard, De Oliveira et al. (2024) focused on the biotechnological and educational potential of Inga edulis Martius, offering an illustrative example of such articulation. Based on an analysis of seeds collected during environmental education activities in the Brazilian Amazon, the authors explicitly recognized the potential of this species for both biotechnological applications and environmental protection.

The study found that I. edulis seeds are particularly attractive because of their nutritional and inorganic composition, including significant concentrations of nitrogen, phosphorus, and other elements, as well as their high protein and carbohydrate content. Similarly, their potential for integration into sustainable forestry systems has been noted, reinforcing their ecological and productive value.

The authors conclude that the production and care of I. edulis seedlings in school settings constitute an environmental education practice with a tangible impact, combining meaningful learning, sustainability, and opportunities for biotechnological development in the Amazon region. This approach demonstrates how environmental education can simultaneously articulate biodiversity conservation, responsible economic development, and strengthen quality education.

Complementarily, the study by Lhota et al. (2024) on The International Education for Conservation program provides additional evidence on the scope of environmental education in relation to environmental SDGs. This program, developed over 14 years, evolved from a local environmental education initiative in Balikpapan Bay, Borneo, after undergoing a systematic critical evaluation process.

Over time, the program became highly valued by teachers, which facilitated its expansion from Kalimantan to other Indonesian islands and even to countries outside of Asia. Its curriculum has been continuously improved through periodic reviews that systematically incorporate the feedback of participating teachers.

Overall, this experience shows that environmental education programs designed with solid pedagogical foundations and continuous evaluation mechanisms can be successfully scaled up, contributing to biodiversity conservation and sustainable development at both regional and international levels.

3.3.4 Impact on Social and Economic SDGs

In addition to its ecological impact, environmental education plays an important role in advancing several social and economic SDGs, such as poverty eradication (SDG 1), food security (SDG 2), promoting health and well-being (SDG 3), gender equality (SDG 5), promoting decent employment and economic growth (SDG 8), and reducing social disparities (SDG 10). These benefits are particularly evident in places where education drives human development and public policies.

In this context, Bynoe (2024) reports a study focused on environmental education at the university level in Guyana and analyzes its role in a developing country. The author notes that environmental education is widely recognized as a key complementary tool for conserving the natural environment, as it strengthens university students' cognitive and critical thinking skills.

From this perspective, teaching has great power to increase knowledge of environmental problems in developing nations, which, in turn, helps make government policies more appropriate and enables clear actions to care for and improve the environment. This highlights the important role of environmental education in creating conditions conducive to informed and responsible decision-making.

Thus, environmental education contributes indirectly to the development of a sustainable economy. This can be achieved by improving our perception of the environment and promoting rules that align with the principles of sustainability (Leal Filho et al., 2020). By developing skills, values, and ways of thinking, this education helps shape key individuals. These individuals are necessary for a long-term transition to more equitable and sustainable economies.

Complementarily, the research by Ayu Setianingrum et al. (2024) provides additional evidence of the social dimensions of environmental education from a cultural and religious perspective. Their study analyzed the values and practices of environmental education from an Islamic perspective, noting that Islam, understood as a comprehensive way of life, establishes clear principles regarding the relationship between humanity and nature, as set out in the Quran and the teachings of the Prophet Muhammad.

The authors analyze the moral foundations of Islamic environmentalism, highlighting ideas such as the human duty to care for creation, the interconnectedness of all living beings, and prudence in the use of resources. This study examined the daily application of these principles in real-world sustainability actions, including water conservation, sustainable agriculture, and ethical consumption.

In short, these findings show that environmental education, when combined with socioeconomic conditions and cultural value systems, can greatly help achieve social and economic SDGs, reinforcing pathways to inclusive and sustainable development.

3.3.5 Measuring Contributions to SDGs

Defining and systematizing the contribution of environmental education to the fulfillment of the SDGs poses methodological challenges that require immediate attention, given that this contribution manifests across educational, social, political, and territorial levels and unfolds at various intervention scales (Biasutti et al., 2016). Consequently, it is essential to have evaluation approaches that do not reduce this phenomenon to isolated indicators but rather allow an understanding of its complexity and the multidimensional nature of the effects of educational actions.

Rosa et al. (2024) focused on analyzing environmental education in schools and offered a critical approach to these challenges. Based on a review of the international literature and research conducted in Brazil, the authors find that a significant proportion of the work focuses on pedagogical practice, followed by approaches to educational policy and teacher training.

However, the study also showed constant problems between the theoretical idea and its application in daily tasks. Similarly, little attention has been paid to active public policies and the structural situations necessary to improve professional training and theoretical development in the sector. Overall, these issues indicate that the calculation of the contribution of environmental education to achieving the SDGs should not be limited to assessing quick effects or detailing personal actions. On the contrary, it requires a more concrete link between educational work, critical thinking, and the creation of theoretical models that explain the difficulties of educational pathways.

From this perspective, the analysis emphasizes the need to take a critical view that challenges conventional, traditional, or overly simplistic ideas about environmental education. Simultaneously, it highlights the value of encouraging reflective pathways for changing school assignments so that environmental education serves as a space for review and change, not just a set of actions without context.

Complementarily, Bustamante Toro and López Castaño (2024) provide innovative methodological tools to document these contributions through social cartography as a teaching strategy in environmental education. Their research experience conceives cartography as a mediating device that facilitates understanding of the interactions between the classroom, school, and territory, integrating the spatial, social, and cultural dimensions of learning.

The results showed that, in almost all cases, students' work considered the natural and cultural sites of their towns and highlighted the most significant environmental problems, such as economic and production issues in the area. This form of participation helps demonstrate how environmental education improves understanding of the relationships among social development, economic activity, and sustainability and provides high-quality evidence to assess its contribution to SDGs.

Overall, these tasks show that accurately assessing the impact of environmental education requires integrated approaches that combine theory, practice, and regional context while recognizing educational actors as people who help build knowledge focused on sustainable development.

3.4 CONCLUSION

3.4.1 Summary of Transformative Potential

The overall analysis of this chapter reveals a key idea: Environmental education is an important tool for driving major changes in response to the sustainability challenges of the 21st century. Its effect is not only to teach people on a personal level but also to help create ways of thinking, strengthen social customs, change organizational systems, and improve group work, thereby changing the relationship between people and the natural world.

The empirical studies reviewed confirm that environmental education can translate into observable changes in pro-environmental behaviors. An illustrative example is the research by Yin et al. (2025), conducted in tourism contexts in China, which showed that the articulation between educational processes and experiences of wonder significantly favors the adoption of environmentally responsible behaviors. However, these effects are enhanced when programs coherently integrate cognitive, emotional, and experiential dimensions, overcoming fragmented or exclusively informative approaches to training.

One of the most significant contributions of environmental education is its ability to transcend individual transformation and drive collective social change. In this sense, Ballard et al. (2024) show that citizen-science initiatives not only strengthen environmental behaviors and management practices but also promote community ties, social cooperation, and forms of collective action. This strengthening of social capital is key to addressing systemic environmental problems, the resolution of which exceeds the capabilities of isolated individuals.

3.4.2 Critical Factors for Effectiveness

The analysis identifies the critical factors that determine the effectiveness of environmental education as a strategic response. First, the integration of multiple scientific, traditional, and experiential knowledge emerges as an indispensable element for building deep and situated understanding. Slikkerveer and Gellaerts (2024) show that transdisciplinary approaches aimed at articulating science and ethnoscience foster the development of citizens with global ecological awareness capable of interpreting the complexity of contemporary environmental problems.

Second, strengthening institutional leadership and organizational capacity is necessary for sustaining and expanding educational impact. Studies such as those by Mishra et al. (2025) highlight the strategic role of school leadership in shaping institutional cultures, mobilizing communities, and consolidating consistent practices. This suggests that lasting transformation does not depend solely on isolated programs or projects but on organizational environments that favor innovation and continuous learning.

The third key factor is the explicit incorporation of equity and environmental justice criteria. Foust et al. (2025) presented evidence indicating that greater attention to these dimensions strengthens the relevance and legitimacy of educational interventions. By integrating perspectives on justice, environmental education can expand its transformative potential by addressing the structural inequalities underlying many processes of environmental degradation, particularly in socially vulnerable contexts.

3.4.3 Contributions to Global Sustainable Development

The review conducted in this chapter confirms that environmental education contributes significantly to various dimensions of sustainable development beyond its direct link to SDG 4 (Quality Education). Its impact extends to environmental, social, and economic goals, demonstrating its cross-cutting nature.

Bradecki et al. (2024) demonstrated how environmental education centers integrate principles of architectural sustainability with innovative educational proposals, yielding simultaneous benefits for education, sustainable urban development, and technological innovation. Complementarily, Purnomo et al. (2024) illustrate how environmental education can be articulated with cultural and religious values, as in the case of sustainable Pesantren, to promote comprehensive community development.

This approach to generating multiple benefits by advancing personal skills, strengthening the collective fabric, promoting innovation, and protecting natural environments positions environmental education as a key strategy with demonstrable benefits across various fields.

3.4.4 Directions for Future Development

Based on this analysis, several priority areas were identified to maximize the transformative potential of environmental education. First, it is necessary to develop methodologies that critically and pedagogically integrate emerging technologies. Immersive virtual reality demonstrates that these tools can amplify educational impact when designed with clear learning objectives and learning process characteristics in mind.

Second, scaling up successful initiatives requires systematic attention to contextual factors and strengthening of institutional capacities. Tiemann and Ring (2025) showed that inequalities in geographic access to environmental education programs can limit their reach, necessitating specific strategies to overcome logistical, economic, and territorial barriers.

Likewise, generating robust evidence of long-term impacts must be prioritized as a research focus. The experience documented by Lhota et al. (2024) shows that systematic monitoring and continuous improvement of environmental education programs can produce sustained effects that transcend local contexts and project themselves onto regional and international scales.

3.4.5 Imperative for Coordinated Action

Overall, the evidence presented makes a strong case for joint action to realize the potential of environmental education as a key response to the challenges of global sustainability. Kopnina (2020) reinforces this imperative, arguing that the effective realization of environmental education's potential requires coordinated action across educational institutions, policy frameworks, and community organizations, moving beyond fragmented initiatives toward systemic integration of sustainability principles throughout educational systems. This action must be taken simultaneously at multiple levels: Improving individuals' capacities, strengthening educational institutions, creating supportive public policies, and building mutual support networks that encourage the sharing of knowledge and resources.

The magnitude and urgency of issues such as climate change, biodiversity loss, and ecosystem degradation require solutions of unprecedented scope and speed to address them. However, the evidence also shows that effective environmental education depends on deliberate methods rooted in local contexts and focused on building lasting commitments over time.

Although environmental education is not a single, self-sufficient solution to all sustainability challenges, the studies reviewed consistently demonstrate that it is an irreplaceable strategy for driving the social, economic, and political transformations needed to move toward a more just and sustainable future. Its full realization depends on sustained investment, ongoing innovation, and a collective commitment to the principles of equity, participation, and responsibility underpinning both quality education and sustainable development.

REFERENCES

Ardoin, N. M., Bowers, A. W., & Gaillard, E. (2020). Environmental education outcomes for conservation: A systematic review. *Biological Conservation, 241*, 108224. https://doi.org/10.1016/j.biocon.2019.108224

Asli, S., Abu-Alhiga, R., Teti, T., Algmal, S., Hofstein, A., Shehadeh-Nasser, A., & Hugerat, M. (2024). How participation in a teachers' eco-pedagogy workshop affects the promotion of teachers' environmental education and organizational concepts. *European Journal of Educational Research, 13*(1), 1–12. https://doi.org/10.12973/eu-jer.13.1.341

Ayu Setianingrum, D., Setiyo, & Dwiyanto, A. (2024). Environmental education through Islamic lens: Values and practices. *E3S Web of Conf., 482*, 04014. https://doi.org/10.1051/e3sconf/202448204014

Ballard, H. L., Lindell, A. J., & Jadallah, C. C. (2024). Environmental education outcomes of community and citizen science: A systematic review of empirical research. *Environmental Education Research, 30*(6), 1007–1040. https://doi.org/10.1080/13504622.2024.2348702

Biasutti, M., De Baz, T., & Alshawa, H. (2016). Assessing the infusion of sustainability principles into university curricula. *Journal of Teacher Education for Sustainability, 18*(2), 21–40. https://doi.org/10.1515/jtes-2016-0012

Blasco-Magraner, J. S., Marín-Liébana, P., Hurtado-Soler, A., & Botella-Nicolás, A. M. (2025). The use of soundscapes in environmental education: Teachers' competencies in auditory analysis and emotional identification. *Behavioral Sciences, 15*(6), 744. https://doi.org/10.3390/bs15060744

Bradecki, T., Uherek-Bradecka, B., Tofiluk, A., Laar, M., & Natanian, J. (2024). Towards sustainable education by design: Evaluating pro-ecological architectural solutions in centers for environmental education. *Sustainability, 16*(12), 5053. https://doi.org/10.3390/su16125053

Bustamante Toro, C. A., & López Castaño, C. E. (2024). The social cartography in key of environmental education to understand the school-territory. *Territorios*, (50), 1–6. https://doi.org/10.12804/revistas.urosario.edu.co/territorios/a.11660

Bynoe, P. (2024). Environmental education at university level in a developing country context: Current trends and future challenges in Guyana. In *A Sourcebook for Environmental Education: A Practical Review Based on the Belgrade Charter* (pp. 52–71). CRC Press. https://doi.org/10.1201/9781003579663-4

Chawla, L. (2020). Childhood nature connection and constructive hope: A review of research on connecting with nature and coping with environmental loss. *People and Nature, 2*(3), 619–642. https://doi.org/10.1002/pan3.10128

Chen, Y., & McCreary, A. (2025). The impacts of environmental education on homeschoolers' attitudes toward nature. *Journal of Adventure Education and Outdoor Learning*, *25*(1), 84–99. https://doi.org/10.1080/14729679.2024.2372395

De Oliveira, A. V., Kersting, M., da Costa, A. B., & de Souza Schneider, R. D. C. (2024). Biotechnological and environmental education potential of inga edulis martius: Seed collected in the amazon-brazil: Potential of Inga edulis seeds. *BIOTROPIA*, *31*(3), 316–329. https://doi.org/10.11598/btb.2024.31.3.2201

Fernández-Llamazares, Á., Lepofsky, D., Lertzman, K., Armstrong, C. G., Brondizio, E. S., Gavin, M. C., Lyver, P. O. B., Nicholas, G. P., Pascua, P. A., Reo, N. J., Reyes-García, V., Turner, N. J., Yletyinen, J., Anderson, E. N., Balée, W., Cariño, J., David-Chavez, D. M., Dunn, C. P., Garnett, S. C., Greening, S., S. J., Kuhnlein, H., Molnár, Z., Odonne, G., Retter, G.-B., Ripple, W. J., Sáfián, L., Bahraman, A. S., Torrents-Ticó, M., & Vaughan, M. B. (2021). Scientists' warning to humanity on threats to indigenous and local knowledge systems. *Journal of Ethnobiology*, *41*(2), 144–169. https://doi.org/10.2993/0278-0771-41.2.144

Foust, C. R., Brannon, D., & Bexell, S. M. (2025). Alignment thinking and complementary curriculum: A qualitative study addressing the urgent need for environmental education. *Journal of Curriculum Studies*, *57*(3), 303–326. https://doi.org/10.1080/00220272.2024.2392599

Fraisl, D., Hager, G., Bedessem, B., Gold, M., Hsing, P.-Y., Danielsen, F., Hitchcock, C. B., Hulbert, J. M., Piera, J., Spiers, H., Thiel, M., & Haklay, M. (2022). Citizen science in environmental and ecological sciences. *Nature Reviews Methods Primers*, *2*(1), 64. https://doi.org/10.1038/s43586-022-00144-4

Frankie, G., & Vinson, B. (2004). Conservation and environmental education in rural Northwestern Costa Rica. *Biodiversity Conservation in Costa Rica*, 247–256. https://doi.org/10.2307/jj.26435419.22

Gilleran Stephens, C., Antwi, S. H., & Linnane, S. (2025). Universal Design for Learning (UDL): A framework for re-design of an Environmental Education (EE) outreach program for a more inclusive and impactful Science Festival event. *Discover Education*, *4*(1), 217. https://doi.org/10.1007/s44217-025-00660-x

Grimmette, K. A. (2014). *The Impacts of Environmental Education on Youth and Their Environmental Awareness*. University of Nebraska-Lincoln. https://digitalcommons.unl.edu/envstudtheses/135

Hankin, C., & Hogarth, H. (2025). Becoming-wild with chalk and paintbrush: Material-multispecies moments for re-imagining environmental education pedagogies. *Australian Journal of Environmental Education*, 1–21. https://doi.org/10.1017/aee.2025.5

Johnson, J. C. (2010). *Seeds of Change: Planting a Path to Peace*. Lee & Low Books. https://doi.org/10.5860/choice.48-2343

Karmakar, G. (2025). Transformative learning with Wangari Maathai: Fostering environmental education and sustainability through the green picturebook seeds of change: Planting a path to peace. *Journal of Human Values*, *31*(1), 97–117. https://doi.org/10.1177/09716858241263129

Kopnina, H. (2020). Education for the future? Critical evaluation of education for sustainable development goals. *The Journal of Environmental Education*, *51*(4), 280–291. https://doi.org/10.1080/00958964.2019.1710444

Kraftl, P., Ambreen, S., Armson, D., Badwan, K., Curtis, E., Pahl, K., & Schofield, J. E. (2025). Starting with trees: Between and beyond environmental education. *British Educational Research Journal*, *51*(2), 782–801. https://doi.org/10.1002/berj.4099

Leal Filho, W., Eustachio, J. H., Caldana, A. C., Will, M., Lange Salvia, A., Rampasso, I. S., Anholon, R., Platje, J., & Kovaleva, M. (2020). Sustainability leadership in higher education institutions: An overview of challenges. *Sustainability, 12*(9), 3761. https://doi.org/10.3390/su12093761

Lhota, S., Bukit, Windi L., Razafindravony, Lovasoa E., & Randrianarivelo, H. (2024). Education for conservation, and evolving international program and 14-year curriculum of environmental education. *Folia Primatologica, 95*(4–6), 451–461. https://doi.org/10.1163/14219980-bja10031

Markkanen, S., & Anger-Kraavi, A. (2019). Social impacts of climate change mitigation policies and their implications for inequality. *Climate Policy, 19*(7), 827–844. https://doi.org/10.1080/14693062.2019.1596873

Massuchin, D. P., Venturi, T., & Iared, V. G. (2025). Educação ambiental e educação em saúde sob uma lente ecofenomenológica: repensando o currículo educacional. *Revista Pesquisa Qualitativa, 13*(34), 403–431. https://doi.org/10.33361/RPQ.2025.v.13.n.34.782

Menton, M., Larrea, C., Latorre, S., Martinez-Alier, J., Peck, M., Temper, L., & Walter, M. (2020). Environmental justice and the SDGs: From synergies to gaps and contradictions. *Sustainability Science, 15*(6), 1621–1636. https://doi.org/10.1007/s11625-020-00789-8

Mishra, A., Mathur, A., & Agnihotri, S. (2025). Sustainable environmental education through school-community collaboration: A case study from India. *International Journal of Environmental Sciences*, 723–729. https://doi.org/10.64252/fcy3xm08

O'Brien, K. (2018). Is the 1.5°C target possible? Exploring the three spheres of transformation. *Current Opinion in Environmental Sustainability, 31*, 153–160. https://doi.org/10.1016/j.cosust.2018.04.010

Purnomo, J., Anantanyu, S., Saptaningtyas, H., & Mangunjaya, F. M. (2024). Prophetic approach in environmental education and community empowerment: A case study of sustainable Pesantren Development. *Revista de Gestao Social e Ambiental, 18*(8), 1–26. https://doi.org/10.24857/rgsa.v18n8-047

Redman, E., Wiek, A., & Redman, A. (2018). Continuing professional development in sustainability education for K-12 teachers: Principles, programme, applications, outlook. *Journal of Education for Sustainable Development, 12*(1), 59–80. https://doi.org/10.1177/2455133318777182

Rosa, M. A., Kauchakje, S., & Fontana, M. I. (2024). Environmental education at school: International literature and analysis of Brazilian studies. *Revista Brasileira de Educação, 29*, e290030. https://doi.org/10.1590/S1413-24782024290030

Rousell, D., & Cutter-Mackenzie-Knowles, A. (2020). A systematic review of climate change education: Giving children and young people a 'voice' and a 'hand' in redressing climate change. *Children's Geographies, 18*(2), 191–208. https://doi.org/10.1080/14733285.2019.1614532

Roy, F. G. (2024). Empowering special education: Fostering environmental education, nature awareness, and sustainable practices. In *Fostering an Ecological Shift Through Effective Environmental Education* (pp. 325–344). IGI Global Scientific Publishing. https://doi.org/10.4018/979-8-3693-2577-3.ch018

Rubáš, D., Matějček, T., & Řezníčková, D. (2024). Sense of place in geography and environmental education: A literature review of empirical research. *GeoScape, 18*(1). https://doi.org/10.2478/geosc-2024-0003

Slikkerveer, L. J., & Gellaerts, S. L. (2024). The indigenous knowledge systems-based ethnoscientific model of Integrated Eco-Education (IEE): Multicultural prospects of environmental education. In T. Kochetkova (Ed.), *Fostering an Ecological Shift Through Effective Environmental Education* (pp. 108–137). IGI Global Scientific Publishing. https://doi.org/10.4018/979-8-3693-2577-3.ch008

Tiemann, A., & Ring, I. (2025). Quantifying and mapping forest-related environmental education and training in Saxony, Germany. *Ecosystem Services, 73*, 101719. https://doi.org/10.1016/j.ecoser.2025.101719

Tillmann, S., Tobin, D., Avison, W., & Gilliland, J. (2018). Mental health benefits of interactions with nature in children and teenagers: A systematic review. *Journal of Epidemiology and Community Health*, *72*(10), 958. https://doi.org/10.1136/jech-2018-210436

Tyler, M.-E. (2025). The environmental education imperative 2024. In *The Routledge Handbook of Global Sustainability Education and Thinking for the 21st Century* (pp. 623–638). Routledge India. https://doi.org/10.4324/9781003171577-50

Wamsler, C., Osberg, G., Osika, W., Herndersson, H., & Mundaca, L. (2021). Linking internal and external transformation for sustainability and climate action: Towards a new research and policy agenda. *Global Environmental Change*, *71*, 102373. https://doi.org/10.1016/j.gloenvcha.2021.102373

Whitburn, J., Linklater, W., & Abrahamse, W. (2020). Meta-analysis of human connection to nature and proenvironmental behavior. *Conservation Biology*, *34*(1), 180–193. https://doi.org/10.1111/cobi.13381

Xie, T., & Yang, Y. (2025). Use of immersive virtual reality in environmental education: Effects on environmental empathy, skill transfer, and attitudes. *Interactive Learning Environments*, *33*(4), 3091–3105. https://doi.org/10.1080/10494820.2024.2436947

Yin, C., Li, W., & Yang, X. (2025). A study of the mechanism of environmental education and awe on tourists' pro-environmental behavior. *International Journal of Tourism Research*, *27*(2), e2798. https://doi.org/10.1002/jtr.2798

Section II

Actors and Ecosystem of Environmental Education

4 Formal Educational Institutions

4.1 PRIMARY AND SECONDARY EDUCATION: CURRICULUM INTEGRATION

4.1.1 Contemporary Challenges in Curriculum Integration

The effective incorporation of environmental education (EE) into primary and secondary school curricula faces various obstacles, ranging from structural constraints to cultural resistance rooted in traditional education systems (Ardoin et al., 2020). Jekel Könnel et al. (2025) examine the effects of an outdoor learning program, "Gewässer Campus," within the framework of EE, highlighting both the opportunities and challenges associated with its integration.

Their quasi-experimental study with 231 German lower and upper secondary school students revealed that learning activities during the project day led to significant knowledge acquisition. More importantly, they found that high individual values were obtained for the test items for preservation and low values for the test items for utilization. These findings suggest that rigorously designed programs can significantly shape students' environmental values. However, they also underscore the need to adjust EE modules within the framework of sustainable development to the characteristics of the group, considering specific teaching and learning needs at the school and institutional levels.

Carbonell-Alcocer and Gertrudix (2025), in their research on teachers' perceptions of effective resources and methodologies for EE for sustainability, provide additional evidence of this implementation gap. Their study of 271 general education teachers in the Community of Madrid found a contradiction between theory and practice, scant mention of current policies, and a lack of working conditions for training and theoretical advancement.

This evidence illustrates a persistent gap between curricular aspirations and implementation realities that affect the effectiveness of EE in formal school contexts. The researchers concluded that continuous training of teachers in active methodologies and in the use of pedagogical-communicative content related to recycling and the circular economy is essential.

4.1.2 Innovative Methodologies for Curriculum Integration

The design of innovative methodologies for curriculum integration has become a strategic priority, as it helps educators overcome the limitations of traditional approaches that treat EE as isolated content rather than as a cross-cutting perspective that can enrich all areas of learning (Wals & Benavot, 2017).

DOI: 10.1201/9781003779667-6

Nacaroğlu and Göktaş (2025) studied the effect of technology-supported guided inquiry EE on future science teachers' attitudes toward the sustainable environment, self-efficacy in EE, and digital literacy, demonstrating promising approaches to this integration. Their study with 33 fourth-year future science teachers used a mixed embedded design and found a significant difference between pre-test and post-test scores obtained from the overall environmental education self-efficacy (EESE) scale.

More specifically, they found a significant difference between the scores obtained from the EESE subdimensions of content knowledge (CK) and instructional strategies (IS) in favor of the post-test. They also reported a significant difference in favor of the post-test between the scores obtained from the general attitudes toward sustainable environment (AtES) scale and the subdimension of environmental resource use (ERU).

These findings suggest that technology-supported guided research approaches may be particularly effective in developing both pedagogical competencies and environmental engagement among future educators. Researchers are organizing technology-supported guided research learning environments for EE.

Basheer et al. (2025) examine the impact of an intervention involving plastics and bioplastics on the green chemistry and sustainability awareness of future science teachers, as well as their attitudes toward EE, thereby providing further evidence for the efficacy of innovative methodologies. Their study of 198 future science teachers found a significant increase in sustainability awareness (1.92–4.36) and green chemistry awareness (2.17–4.20) among future teachers. Paired sample t-tests confirmed significant improvements in green chemistry awareness, sustainability awareness, and attitudes toward EE.

4.1.3 Integration of Emerging Technologies

Emerging technologies offer unprecedented opportunities to transform the curricular integration of EE, providing tools that can make complex environmental concepts more accessible, create immersive experiences that connect students with distant ecosystems, and facilitate collaboration among students from diverse geographic regions (Rousell & Cutter-Mackenzie-Knowles, 2020).

Nguyen et al. (2025) demonstrate the transformative potential of chatbots in EE through their value-sensitive design (VSD). Their study applied VSD, involving empirical, conceptual, and technical research, to center human values in the development and evaluation of large language model (LLM)-based chatbots within a middle school environmental science curriculum.

The findings illuminated considerations for chatbots to support student identity development, well-being, human–chatbot relationships, and environmental sustainability. The researchers mapped the emerging values from VSD research to various stages of GenAI technology development: Conceptualization, development, and evaluation. This study demonstrates how to conduct contextual, value-sensitive research on emerging artificial intelligence (AI) technologies in educational settings.

Gesthuizen et al. (2025) provide additional evidence of the potential of immersive technologies in their study on building understanding of sustainability and EE within a virtual world. Their study used the "Teaching to Inspire" the UN Sustainable Development Goals and the socioecological systems framework.

The results highlight the importance of curiosity, passion, and drive in the conative domain in encouraging student engagement and deepening their understanding of sustainability. The study also underscored the critical role of teachers in promoting critical thinking, inquiry, and reflection in virtual environments to build a sense of place and strengthen emotional connections to the topics addressed.

4.1.4 Development of Specialized Teaching Skills

The development of specialized teaching skills in EE is a fundamental prerequisite for effective curriculum integration in higher education. Recent research indicates that a significant proportion of teachers lack the necessary training to implement EE effectively in schools. Consequently, comprehensive professional development programs are needed to strengthen both mastery of environmental content and the use of specialized teaching methods (Redman et al., 2018).

Grillia and Daskolia (2025) in "Why am I involved in environmental education?" used narrative research to explore the identity and practices of teachers committed to EE. Their study was based on self-reflective narratives from four Greek primary school teachers who were recognized for their personal commitment to EE.

The study used the ontological and epistemological principles of narrative inquiry by Clandinin (2006) to "examine teachers' reflections on why they became involved in environmental education." The researchers argued that "involving teachers in generative reflective processes contributes to gaining insights into environmental education 'from within' and 'from below,' while allowing us to rethink and reimagine research in the field."

Shin and van Riper (2025) provide evidence of the importance of early formative experiences for the development of teaching competencies in their study on how childhood outdoor recreation and EE shape pro-environmental behavior among adults. Their study, using data from an online panel survey of respondents in Illinois, USA (n = 617), found that engagement in outdoor recreation during childhood positively influenced connection to nature in adulthood, which in turn reinforced environmental values as guiding principles in life.

Significantly, childhood EE experiences did not predict a connection with nature in adulthood, but they did have a positive influence on environmental values. This finding suggests that different types of training experiences can contribute to teaching skills through different channels, with important implications for designing professional development programs.

4.1.5 Assessment of Environmental Learning in Formal Contexts

The development of appropriate assessment systems for environmental learning in formal educational contexts presents unique challenges that go beyond traditional measures of factual knowledge to include assessing complex competencies, such as systems thinking, environmental problem-solving, and commitment to sustainability values (van de Wetering et al., 2022).

Luan et al. (2025), in their exploration of college students' conceptions of EE through phenomenographic analysis, provide important insights into these assessment challenges. Their study with 36 college students identified five qualitatively different categories of conceptions of EE, namely, receiving information, solving problems, and acting.

These categories were ordered hierarchically from lowest to highest, reflecting increasing complexity: the first two were categorized as passive conceptions, and the latter three as active conceptions. The findings showed that passive and active conceptions were evenly distributed among students.

Crucially, a χ^2 analysis revealed an observable correlation between students' conceptions and their adopted learning strategies, where students with more passive conceptions tended to employ surface learning strategies, while those with more active conceptions leaned toward deep learning strategies.

4.1.6 Successful Cases of Curriculum Integration

Documenting successful cases of curriculum integration provides concrete evidence of effective strategies and factors that facilitate the successful implementation of EE in formal educational systems.

The research conducted by Salazar et al. (2024) presents compelling evidence regarding the impact of EE on pro-environmental behavior among children in Chile, specifically in the context of plastic waste management. Their study used a multi-value treatment effects model to evaluate the effects of a green school certification program on the pro-environmental behavior of Chilean children.

The results showed a positive effect of higher school certification levels on children's behavior toward plastic waste. This effect appeared to be stronger in practices where children have more decision-making power (e.g., packing a lunch box). The researchers observed that the inverse effect observed when moving from the basic to the intermediate level of certification is in line with the potentially nonlinear effects of EE on pro-environmental behavior regarding plastic consumption and disposal.

4.2 HIGHER EDUCATION: RESEARCH AND LEADERSHIP TRAINING

4.2.1 Transforming University Research

Higher education institutions occupy a unique position in the EE ecosystem, functioning simultaneously as knowledge generation centers, training grounds for future leaders, and catalysts for innovation in sustainability. The transformation of university research toward more integrated, sustainability-oriented approaches represents both an opportunity and a fundamental challenge in maximizing the impact of these institutions (Leal Filho et al., 2020).

These results are clearly related to the study by Ouellet Dallaire et al. (2018), entitled "Creating spaces for sustainability literacy: the case of student-centered symposia." The study demonstrates that a student-driven initiative, such as a student-centered symposium held periodically at the university, can open and consolidate educational spaces for students. This promotes sustainability learning by linking student participation with knowledge production and academic and institutional life (Ramísio et al., 2019).

From this perspective, the transformative potential of student groups cannot be explained solely by personal motivation; it also depends on universities enabling and sustaining institutional platforms, such as symposia, networks, and formal mechanisms for participation, which provide continuity and legitimacy to these initiatives.

Furthermore, having clear guidelines and resources is crucial for the continuity of such initiatives. Therefore, aspects such as succession planning and leadership must remain effective over time (Ouellet Dallaire et al., 2018). This reinforces the importance of universities supporting these actions so that they can grow and consolidate in the field of EE. Azarkhin and Ivanova (2024) provide conceptual frameworks for this transformation through the organizational and legal principles of EE in higher education institutions. Their research concludes that "environmental education is actively carried out at all educational stages and that each educational level is complete and serves as a springboard for the formation of ecological awareness and thinking."

The researchers developed the model of EE in higher education institutions, which includes objective, methodological, and methodology blocks. The purpose of EE is defined as the establishment of the relationship between the future subject of professional activity and the environment.

4.2.2 Development of Interdisciplinary Programs

The development of interdisciplinary programs in higher education represents a fundamental evolution from traditional discipline-based academic structures toward more integrated approaches that can address the complexity of contemporary environmental challenges. This transformation requires curricular changes and modifications to organizational structures, assessment systems, and institutional cultures (Tejedor et al., 2019).

Hu et al. (2025) on EE in low- and middle-income countries (LMICs) provide evidence on the effectiveness of interdisciplinary programs through a robust meta-analysis. Their study analyzed 187 independent effect sizes, involving 34,283 participants, and found a positive and significant effect of EE in (LMICs). Specifically, participation in EE programs was associated with increased environmental knowledge, attitudes, and behaviors.

Moderator analyses revealed that the results differed by intervention duration, measurement time, age, and national development level, whereas study design, education level, intervention type, and gender did not. This underscores the importance of designing interdisciplinary programs that consider specific contextual factors while maintaining consistency in core objectives.

4.2.3 Training Environmental Leaders

Training environmental leaders in higher education institutions goes beyond the simple transmission of technical knowledge to include the development of leadership, communication, and systemic thinking skills, as well as the facilitation of organizational and social change. This training requires innovative pedagogical approaches that integrate experiential learning with the development of analytical and reflective skills (Evans, 2019).

Dean and Gilbert (2025) highlight the significance of observing individuals' autonomous actions, emphasizing that shared agency in school-based outdoor EE offers valuable insights into methodologies for leadership development. Their interpretive phenomenological study interviewed 12 participants who regularly implemented outdoor EE in a K-12 setting.

The researchers analyzed the participants' narratives using a whole-part-whole approach, identifying five key themes that highlight how teachers construct and curate outdoor learning experiences that foster student agency. The study found that navigating shared agency among teachers, students, and places is a complex and ongoing process that disrupts the traditional classroom hierarchy.

Without a doubt, the data show that outdoor EE can be a useful way to improve student independence by placing them at the center of their own learning, encouraging ongoing action, and supporting equitable education. Therefore, ideas grounded in shared authority seem to be highly effective for developing environmental leaders with technical expertise and a clear sense of social responsibility.

4.2.4 Action-Oriented Research

Action-oriented research in higher education represents an evolution toward more participatory and socially engaged approaches that connect academic knowledge generation with the solution of real environmental problems in local and regional communities.

Zhao et al. (2024) improved pro-environmental behavior through nature-based EE as an example of action-oriented research. Their empirical analysis, based on a randomized controlled trial design, tested the effects of nature-based EE on pro-environmental behaviors.

The results indicate that EE involving contact with nature significantly improves participants' pro-environmental behavior. More importantly, this study found that EE involving contact with nature improves pro-environmental behavior mainly by improving environmental knowledge, attitudes, and social norms.

This study demonstrates how university studies can generate rigorous evidence of the effectiveness of educational interventions while also providing practical tools for educators and policymakers to design more effective programs.

4.3 TECHNICAL AND VOCATIONAL TRAINING IN ENVIRONMENTAL SKILLS

4.3.1 Development of Specialized Technical Curricula

Technical and vocational training in environmental skills has emerged as a critical component of the educational ecosystem, responding to the growing demand for professionals who can implement environmental solutions in the industrial, government, and service sectors. This training requires the development of specialized curricula that integrate technical knowledge with an understanding of complex environmental systems, as well as communication and leadership skills (Vona et al., 2018).

Shirisha et al. (2025) discuss the significance of cross-sectoral approaches in professional training, particularly in the context of EE and strategic communication in rural India. Their project analyzed how communication strategies are used in EE in rural India to determine how different groups can work together to promote sustainability awareness.

The researchers identified schools, nongovernmental organizations (NGOs), local self-government bodies (panchayats), and the media as central stakeholders and investigated how they provide environmental knowledge. The findings revealed that both traditional forms of communication and newer school activities are important steps in sharing information.

Crucially, this type of communication approach has been shown to greatly impact people's behavior and opinions toward water conservation, waste management, and biodiversity protection. This suggests that effective technical training should include skills in strategic communication and community engagement, in addition to specialized technical expertise.

4.3.2 Professional Certification Programs

Professional certification programs in environmental competencies have proliferated to address the need for certified professionals to ensure the effective implementation of environmental regulations, sustainable technologies, and environmental management practices in diverse organizations.

The study conducted by Vaca-León et al. (2025) examines the role of reptiles in ecological parks dedicated to conservation and EE in Mexico, highlighting the integration of professional education with practical conservation efforts. Their chapter presents the practice, experience, and impact of interactions between reptiles and tourism, highlighting aspects of bioethics, conservation, and EE through the Draconis Mx Herpetarium in the Zacango Ecological Park in Mexico.

This case demonstrates how professional training programs can integrate specialized scientific knowledge with practical applications in public education and conservation efforts. The authors emphasize that discussing the interaction between these vertebrates and humans in ecological parks allows us to understand the magnitude of their presence in life, culture, and perceptions.

The research suggests that effective certification programs should include components that address not only technical skills but also the ethical, cultural, and communicative dimensions of professional environmental work.

4.3.3 Industry-Academia Collaborations

Collaborations between industry and academia in environmental technical training are a key strategy for ensuring that educational programs align with labor market needs and promote innovation in environmental technologies and practices.

Matos et al. (2024) illustrated collaborative approaches to training parataxonomists through an educational trail in the Açu National Forest. Their project developed an EE trail for visitors, seed collectors, parataxonomists, and researchers. The study identified 13 species of Leguminosae and labeled 34 individuals, in addition to 58 individuals belonging to 13 different families.

The researchers defined three trails: (1) Legume trail, (2) Catingueira-cinzenta trail, and (3) Caatinga immersion trail. The authors and visitors tested these trails between July 23 and 24, 2022, and concluded that they are an important tool for protecting these environments and raising awareness of local species of flora and fauna.

This example demonstrates how collaborations can generate educational tools that serve multiple purposes: Professional training for parataxonomists, public education for visitors, and the practical conservation of local ecosystems.

4.3.4 Continuing Education and Professional Development

The rapid pace of technological and regulatory change in the environmental field requires robust continuing education systems to keep professionals updated on emerging developments in science, technology, and environmental policy (Brundiers et al., 2021).

The research by da Silva et al. (2024) on conservation units, EE, and empowerment provides insights into continuing education systems. Their case study of Pau Brasil National Park analyzed possible connections between the establishment of a socio-environmental council, the Thematic Chamber of Environmental Education (CTEA) within the PNPB Advisory Council, and the variation in the power to act of this collective and those involved in its creation and operation.

Researchers found that the creation of the CTEA resulted in an increase in the group's power to act, but many challenges remain. They identified a direct relationship between participation, power to act, and positive results. This suggests that effective continuing education systems should include components that strengthen the capacity for participation in environmental governance and collective action.

4.4 SUCCESSFUL CASES OF INSTITUTIONAL TRANSFORMATION

4.4.1 Systemic Transformation in Universities

The systemic transformation of universities toward sustainability represents one of the most promising developments in institutional EE, demonstrating how large and complex institutions can reorganize their operations, curricula, and organizational cultures to align with sustainable principles.

Justin Hougham et al. (2024) reported declines in diversity, equity, and inclusion (DEI) practices within EE organizations, offering important insights into the challenges of institutional transformation. Their research focused on measures reported by EE organizations in Wisconsin and surrounding regions in their work to address DEI regarding inclusive practices, instructional materials, and general EE pedagogy.

The findings demonstrate the need for equity in EE, specifically to support students with diverse learning needs or from underrepresented backgrounds. However, the importance of DEI initiatives versus allocated resources is disjointed, and the lack of consistent demographic data collection leads EE organizations to fall short of DEI action.

This evidence underscores that successful institutional transformation requires not only statements of intent but also adequate resource allocation and monitoring systems to ensure the effective implementation of equity and inclusion policies.

4.4.2 Sustainable Campus Models

Sustainable campus models have emerged as living laboratories where educational institutions can demonstrate sustainability principles while providing experiential learning opportunities for students, faculty, and staff (Ávila et al., 2017; Findler et al., 2019).

Aswita et al. (2024), in their work, "Improving Environmental Literacy in Aceh Tourism through an Ethnoecotourism-Based Environmental Education Model," elucidate integrated strategies for achieving institutional sustainability. Their study examined the effectiveness of an ethnoecotourism-based EE model for tourism actors in the Aceh Province. The model was developed to increase environmental literacy in marine tourism management.

Trials of the model were conducted in the Marine Tourism Area of Sabang City, and it was found that this model could improve the environmental literacy skills of tourism stakeholders in the research location. This was demonstrated by the significant value obtained in $p = 0.000$ (< 0.05), where there was a difference in student scores after receiving the learning process, with the average post-test score of 12.43 being higher than the pre-test score of 8.40.

4.4.3 Strategic Partnerships for Transformation

Strategic partnerships between educational institutions and external organizations have emerged as crucial mechanisms for accelerating institutional transformation by providing resources, expertise, and external accountability to complement internal institutional capacities.

Ross (2025) examines these partnerships within the context of EE center policy through a historical perspective, as detailed in their book. The analysis traces the evolution of these centers, highlighting the influence of the 1970s' rise in socially conscious governance, the public managerialism of the 1980s–1990s, and the increasingly pervasive influence of capitalism.

Ross (2025) also emphasizes that the complexity of education as a socio-political endeavor is made explicit and examines issues including the changing dynamics of environmental stewardship, bureaucratic and political maneuvering, and the effects of activism and globalization. This historical perspective provides important insights into how strategic partnerships must navigate changing political and economic contexts to maintain their long-term effectiveness.

4.4.4 Indicators of Success in Institutional Transformation

Developing appropriate indicators to measure success in institutional transformation presents unique challenges because the most significant changes often unfold over extended periods and may manifest in ways that are difficult to quantify using traditional metrics.

Various studies suggest that institutional change is best assessed by combining quantitative and qualitative data (Disterheft et al., 2015). This comprehensive view helps evaluate rule modernization, fund delivery, plan initiation, and changes in organizational culture. Furthermore, these references should not focus solely on outcomes but also on processes: Structural change is fluid and ongoing, so it requires regular monitoring and robust adjustments.

More recently, Olsson et al. (2022) demonstrated through longitudinal research that institutional transformation in sustainability education is best assessed through multidimensional frameworks that track changes across cognitive, affective, and behavioral domains over extended time periods, confirming the need for assessment approaches that capture both outcomes and the ongoing processes of institutional change.

4.5 CONCLUSION

4.5.1 Summary of Transformative Potential

A positive response to EE in schools is not achieved by introducing isolated topics but rather through a complete change in teaching methods. Evidence shows that the use of new methods, together with institutional support, can lead to clear improvements in how issues are valued and addressed. Unlike traditional methods, this approach helps students learn concepts that endure longer, which support future environmental action.

In line with this idea, Hu et al. (2025) declare that introducing new technologies on virtual sites provides unprecedented opportunities to expand education. Nevertheless, their introduction should be grounded in strong pedagogical principles and clear ethical guidelines so that technology supports instruction rather than merely serving as a new tool (Nguyen et al., 2025).

4.5.2 Critical Factors for Institutional Success

The analysis identified several critical factors that determine the success of EE initiatives in formal institutions. First, the development of specialized teaching skills is fundamental, as illustrated by Grillia and Daskolia (2025) on the identity and practice of committed teachers. The most effective educators are those who combine technical expertise with a deep understanding of appropriate pedagogical methodologies and personal commitment to environmental values.

Second, effective curriculum integration requires more than simply adding environmental courses; it demands the reorganization of academic structures to facilitate interdisciplinary approaches and authentic learning experiences. Luan et al. (2025) report on nature-based EE illustrates how students with more active conceptions of EE tended to adopt deeper learning strategies, suggesting that curricula should be designed to cultivate active engagement rather than passive information reception.

Third, systematic institutional support is essential for sustaining EE initiatives in the long term. Ouellet Dallaire et al. (2018) identified teamwork, commitment, succession development, and leadership as key factors for organizational sustainability. These findings suggest that institutions should invest in their organizational capacity.

4.5.3 Directions for Future Development

The analysis suggests several priority directions for the future development of EE in formal institutions in the region. First, action-oriented research should be expanded to generate more robust evidence of the effectiveness of different pedagogical approaches in diverse contexts. A study on nature-based EE illustrates how rigorous experimental methods can generate practical evidence of the causal mechanisms underlying behavioral changes.

The improvement of environmental programs depends on two critical factors: Evaluation sophistication and external cooperation. It is necessary to implement monitoring systems that measure both the instantaneous effect and persistence of knowledge over time. In addition, synergy with external strategic partners compensates for institutional shortcomings, providing added value that is only possible through collaboration. Such agreements must be designed in accordance with sustainability and shared gain criteria to ensure their success.

4.5.4 Imperatives for Systemic Transformation

The evidence examined here points to the urgent need for a comprehensive transformation of current educational institutions, going beyond mere improvements and generating profound changes in their organizational structures, curricula, and connections with the community at large (Sterling et al., 2018). Above all, this improvement requires strong leadership capable of fostering institutional resilience, building consensus among various communities (administration, faculty, students, and staff), and managing the tensions inherent in any structural change.

Similarly, it is crucial to cultivate skills that do not depend on passing media fads or short-lived funding. This investment extends beyond financial consideration; it also entails fostering ecological values that serve as moral guides, particularly during periods of financial uncertainty and strategic change.

Finally, it is essential to recognize that institutional change does not occur in isolation from other changes. Schools and universities must be integrated into broader educational systems and collaborate with public sectors, private enterprises, and community organizations. This external connection must be forged with equality and commitment to social justice, prioritizing the common good over individual interests.

REFERENCES

Ardoin, N. M., Bowers, A. W., & Gaillard, E. (2020). Environmental education outcomes for conservation: A systematic review. *Biological Conservation*, *241*, 108224. https://doi.org/10.1016/j.biocon.2019.108224

Aswita, D., Noviyanti, A., Apriana, E., Jalaluddin, J., Febriyanti, D., Herlina, H., & Hasmita, I. (2024). Enhancing environmental literacy in Aceh's tourism through an ethnoecotourism based environmental education model. *Jurnal Ilmiah Peuradeun*, *12*(3), 1051–1070. https://doi.org/10.26811/peuradeun.v12i3.1415

Ávila, L. V., Leal Filho, W., Brandli, L., Macgregor, C. J., Molthan-Hill, P., Özuyar, P. G., & Moreira, R. M. (2017). Barriers to innovation and sustainability at universities around the world. *Journal of Cleaner Production*, *164*, 1268–1278. https://doi.org/10.1016/j.jclepro.2017.07.025

Azarkhin, A., & Ivanova, S. (2024). Organizational and legal principles of environmental education in higher education institutions. *E3S Web of Conf.*, *531*, 05031. https://doi.org/10.1051/e3sconf/202453105031

Basheer, A., Gulacar, O., Sindiani, A., & Eilks, I. (2025). The impact of an intervention on plastics and bioplastics on pre-service science teachers' green chemistry and sustainability awareness and their attitudes toward environmental education. *Education Sciences*, *15*(3), 322. https://doi.org/10.3390/educsci15030322

Brundiers, K., Barth, M., Cebrián, G., Cohen, M., Diaz, L., Doucette-Remington, S., Dripps, W., Habron, G., Harré, N., Jarchow, M., Losch, K., Michel, J., Mochizuki, Y., Rieckmann, M., Parnell, R., Walker, P., & Zint, M. (2021). Key competencies in sustainability in higher education—toward an agreed-upon reference framework. *Sustainability Science*, *16*(1), 13–29. https://doi.org/10.1007/s11625-020-00838-2

Carbonell-Alcocer, A., & Gertrudix, M. (2025). Percepción Docente sobre los Recursos y Metodologías Eficaces para la Educación Ambiental para la Sostenibilidad. *Revista Internacional de Educación para la Justicia Social*, *14*(1). https://doi.org/10.15366/riejs2025.14.1.008

Clandinin, D. J. (2006). Narrative inquiry: A methodology for studying lived experience. *Research Studies in Music Education*, *27*(1), 44–54. https://doi.org/10.1177/1321103X060270010301

Dean, S., & Gilbert, A. (2025). 'Watching what they can do on their own is beautiful': A navigation of shared agency in school-based outdoor environmental education. *Journal of Adventure Education and Outdoor Learning*, *25*(2), 563–579. https://doi.org/10.1080/14729679.2024.2314605

Disterheft, A., Caeiro, S., Azeiteiro, U. M., & Filho, W. L. (2015). Sustainable universities: A study of critical success factors for participatory approaches. *Journal of Cleaner Production*, *106*, 11–21. https://doi.org/10.1016/j.jclepro.2014.01.030

Evans, T. L. (2019). Competencies and pedagogies for sustainability education: A roadmap for sustainability studies program development in colleges and universities. *Sustainability*, *11*(19), 5526. https://www.mdpi.com/2071-1050/11/19/5526

Findler, F., Schönherr, N., Lozano, R., Reider, D., & Martinuzzi, A. (2019). The impacts of higher education institutions on sustainable development: A review and conceptualization. *International Journal of Sustainability in Higher Education*, *20*(1), 23–38. https://doi.org/10.1108/ijshe-07-2017-0114

Gesthuizen, R., Tan, H., & Kidman, G. (2025). Inspired to build understanding: Learning about sustainability and environmental education from within a virtual world. *International Research in Geographical and Environmental Education*, *34*(1), 25–42. https://doi.org/10.1080/10382046.2024.2348265

Grillia, N., & Daskolia, M. (2025). 'Why am I involved in environmental education?' Using narrative inquiry to explore the identity and practice of teachers committed to environmental education. *Environmental Education Research*, *31*(1), 29–45. https://doi.org/10.1080/13504622.2024.2309582

Hu, B., Wang, C., Xia, Z., Lu, C., & Ju, X. (2025). Environmental education in low-income and middle-income countries: A systematic review and meta-analysis. *Journal of Environmental Psychology*, *104*, 102613. https://doi.org/10.1016/j.jenvp.2025.102613

Jekel Könnel, E., Geuer, L., Schlindwein, A., Perret, S., & Ulber, R. (2025). The effects of an outdoor learning program, 'GewässerCampus', in the context of environmental education. *Education Sciences*, *15*(5), 550. https://doi.org/10.3390/educsci15050550

Justin Hougham, R., Burgess, S., & Bauer, J. (2024). Reported decreases in diversity, equity, and inclusion practices in environmental education organizations. *Journal of Experiential Education*, *47*(2), 238–243. https://doi.org/10.1177/10538259241232323

Leal Filho, W., Eustachio, J. H., Caldana, A. C., Will, M., Lange Salvia, A., Rampasso, I. S., Anholon, R., Platje, J., & Kovaleva, M. (2020). Sustainability leadership in higher education institutions: An overview of challenges. *Sustainability*, *12*(9), 3761. https://doi.org/10.3390/su12093761

Luan, H., Syu, Y.-L., Lee, M.-H., & Tsai, C.-C. (2025). Exploring undergraduate students' conceptions of environmental education through phenomenographic analysis. *Science Education*, *109*(3), 876–892. https://doi.org/10.1002/sce.21948

Matos, S. S. D., Ganade, G., & Versieux, L. D. M. (2024). Trilha educativa na Floresta Nacional de Açu: Leguminosas como ferramenta de Educação Ambiental e formação de parataxonomistas1. *Hoehnea*, *51*, e592022. https://doi.org/10.1590/2236-8906e592022

Nacaroğlu, O., & Göktaş, O. (2025). The effect of technology supported guided inquiry environmental education on preservice science teachers' attitudes towards sustainable environment, environmental education self-efficacy and digital literacy. *Journal of Science Education and Technology*, *34*(2), 298–313. https://doi.org/10.1007/s10956-024-10188-8

Nguyen, H., Nguyen, V., Ludovise, S., & Santagata, R. (2025). Value-sensitive design of chatbots in environmental education: Supporting identity, connectedness, well-being and sustainability. *British Journal of Educational Technology*, *56*(4), 1370–1390. https://doi.org/10.1111/bjet.13568

Olsson, D., Gericke, N., & Boeve-de Pauw, J. (2022). The effectiveness of education for sustainable development revisited: A longitudinal study on secondary students' action competence for sustainability. *Environmental Education Research*, *28*(3), 405–429. https://doi.org/10.1080/13504622.2022.2033170

Ouellet Dallaire, C., Trincsi, K., Ward, M. K., Harris, L. I., Jarvis, L., Dryden, R. L., & MacDonald, G. K. (2018). Creating space for sustainability literacy: The case of student-centered symposia. *International Journal of Sustainability in Higher Education*, *19*(4), 839–855. https://doi.org/10.1108/IJSHE-08-2017-0126

Ramísio, P. J., Pinto, L. M. C., Gouveia, N., Costa, H., & Arezes, D. (2019). Sustainability strategy in higher education institutions: Lessons learned from a nine-year case study. *Journal of Cleaner Production*, *222*, 300–309. https://doi.org/10.1016/j.jclepro.2019.02.257

Redman, E., Wiek, A., & Redman, A. (2018). Continuing professional development in sustainability education for K-12 teachers: Principles, programme, applications, outlook. *Journal of Education for Sustainable Development*, *12*(1), 59–80. https://doi.org/10.1177/2455133318777182

Ross, A. M. (2025). *The Politics of Environmental Education Centres* (Vol. 1). Springer. https://doi.org/10.1007/978-3-031-82567-5

Rousell, D., & Cutter-Mackenzie-Knowles, A. (2020). A systematic review of climate change education: Giving children and young people a 'voice' and a 'hand' in redressing climate change. *Children's Geographies*, *18*(2), 191–208. https://doi.org/10.1080/14733285.2019.1614532

Salazar, C., Jaime, M., Leiva, M., & González, N. (2024). Environmental education and children's pro-environmental behavior on plastic waste: Evidence from the green school certification program in Chile. *International Journal of Educational Development*, *109*, 103106. https://doi.org/10.1016/j.ijedudev.2024.103106

Shin, S., & van Riper, C. J. (2025). Childhood outdoor recreation and environmental education shape pro-environmental behavior among adults: A study of residents in Illinois, USA. *Society & Natural Resources*, *38*(6), 563–586. https://doi.org/10.1080/08941920.2024.2449048

Shirisha, D., Vijayalakshmi, A., Mukherjee, F., & Emimal, M. (2025). Environmental education and strategic communication in rural India: A cross-sectoral approach. *International Journal of Environmental Sciences*, *11*(4s), 1240–1245. https://doi.org/10.64252/2sa8yr26

Silva, G. A. L. D., Marcolin, C. D. R., & Costa-Pinto, A. B. (2024). Unidad de Conservación, Educación Ambiental Y Poder de Acción:¿ Hacia Sociedades Sostenibles? *Ambiente & Sociedade*, *27*, e00034. https://doi.org/10.1590/1809-4422asoc00341vu27L5OA

Sterling, S., Dawson, J., & Warwick, P. (2018). Transforming sustainability education at the creative edge of the mainstream: A case study of schumacher college. *Journal of Transformative Education*, *16*(4), 323–343. https://doi.org/10.1177/1541344618784375

Tejedor, G., Segalàs, J., Barrón, Á., Fernández-Morilla, M., Fuertes, M. T., Ruiz-Morales, J., Gutiérrez, I., García-González, E., Aramburuzabala, P., & Hernández, À. (2019). Didactic strategies to promote competencies in sustainability. *Sustainability*, *11*(7), 2086. https://doi.org/10.3390/su11072086

Vaca-León, O. I. M., Rojas, E. L., & Ortiz Lara, D. I. (2025). Reptiles in ecological parks aiming for conservation and environmental education in Mexico. In *Human-Animal Relations in Tourism, Leisure and Development: Perspectives from Latin America* (pp. 61–69). CABI GB. https://doi.org/10.1079/9781800626829.0006

van de Wetering, J., Leijten, P., Spitzer, J., & Thomaes, S. (2022). Does environmental education benefit environmental outcomes in children and adolescents? A meta-analysis. *Journal of Environmental Psychology*, *81*, 101782. https://doi.org/10.1016/j.jenvp.2022.101782

Vona, F., Marin, G., Consoli, D., & Popp, D. (2018). Environmental regulation and green skills: An empirical exploration. *Journal of the Association of Environmental and Resource Economists*, *5*(4), 713–753. https://doi.org/10.1086/698859

Wals, A. E. J., & Benavot, A. (2017). Can we meet the sustainability challenges? The role of education and lifelong learning. *European Journal of Education*, *52*(4), 404–413. https://doi.org/10.1111/ejed.12250

Zhao, Y., Liu, X., & Han, X. (2024, December). Enhancing pro-environmental behavior through nature-contact environmental education: An empirical analysis based on randomized controlled experiment design. *Frontiers in Environmental Science*, *12*, 1–15. https://doi.org/10.3389/fenvs.2024.1491780

5 Civil Society Actors

5.1 NONGOVERNMENTAL ORGANIZATIONS: ROLE AND STRATEGIES

Nongovernmental organizations (NGOs) have been key actors in strengthening and disseminating environmental education internationally (Ardoin et al., 2020). Since the 1970s, their presence has become particularly relevant in raising ecological awareness, as they have promoted educational initiatives that have influenced not only the development of public policies but also the ways in which communities adopt and transform their daily practices. Thanks to their independence and direct fieldwork, these organizations have been able to reach populations and territories that formal educational systems often overlook, developing context-sensitive approaches that respond to local environmental realities while contributing to broader global awareness. Their capacity to operate across multiple scales—from grassroots community interventions to international advocacy campaigns—positions them as indispensable bridges between scientific knowledge, policy frameworks, and the everyday environmental practices of diverse communities.

5.1.1 Distinctive Features of Environmental NGOs

NGOs focused on environmental education are distinguished by their flexible organizational models, strong responsiveness to sudden environmental issues, and close contact with populations directly affected by them. These aspects facilitate the development and implementation of adapted educational ideas tailored to real-world contexts that function effectively in local, regional, or international settings.

Another characteristic is their focus on nonformal and informal education, which greatly expands the reach of their educational work to people who typically lack access to mainstream education systems. Through activities such as neighborhood workshops, information projects, training plans, and outreach events, these organizations have helped more people access environmental education, bringing it to the everyday places where people live and work.

The study by Cândido Vendrasco and Guerrero (2025) regarding the role of nonformal teachers in environmental education provides important evidence; conducted in Brazil and Chile, it showed that teachers working in aquariums and zoos seek to elicit emotional responses associated with biodiversity loss and showcase conservation actions aimed at reducing species extinction. In botanical studies, the primary focus is on advancing knowledge of plants and ecosystems, whereas in national parks, the emphasis is on critical thinking.

DOI: 10.1201/9781003779667-7

5.1.2 Methodological Innovation in Environmental Education

Environmental NGOs are distinguished by teaching methods that prioritize active participation, close engagement with the environment, and the creation of shared knowledge (Wals & Benavot, 2017). An important example is popular environmental education, which combines critical pedagogical ideas with technical knowledge of the environment, encouraging learning without hierarchies in which participants help to create knowledge.

In addition, participatory action has become central to these organizations' teaching methods, enabling the integration of learning and social change. With this idea, communities not only receive data on environmental problems in their area but also learn to identify, understand, and address them and to create and apply solutions grounded in the knowledge and experience of their lands.

For example, the effectiveness of these ideas is evident by Wardhani et al. (2024), who examined an environmental education model grounded in the local wisdom of the Dayak Paramasan tribe in Indonesia. The data they obtained showed an important relationship among indigenous values, experiential knowledge, and community unity in environmental education, and that traditional practices, such as agriculture and hunting, help care for the natural environment.

Just as these studies are based on analyses of environmental education, there are social initiative organizations that have created place-based environmental education programs, using local environments as learning sites and building strong ties to the environment, which makes people more committed to caring for and using natural resources wisely.

5.1.3 Collaborative Networks and Strategic Alliances

One of the greatest strengths of NGOs focused on environmental issues is their ability to form and sustain collaborative networks that expand the reach and impact of their educational initiatives. Over the years, these organizations have formed key alliances (Mogren et al., 2019). They have partnered with schools, government agencies, global groups, and industries. This has created a network of mutual support. These networks optimize the use of existing resources, personnel, and technologies. These alliances have simplified the sharing of ideas and training data. This has occurred between groups working in different locations and cultures in the past.

In this way, environmental education can be improved. This helps the methods to work more effectively. It also facilitates the adaptation of ideas when necessary. This is evident in organizations with diverse social and geographical realities. A clear example of plan creation and adaptation can be seen in the study by Repetto et al. (2024), who examined the use of online escape games in environmental education.

This finding supports the idea that virtual learning is beneficial to students. This suggests that teaching methods should be improved to ensure their effectiveness for all types of students. In particular, it allows researchers to identify which aspects of the activity are important, such as its format, accuracy, or perceived difficulty. This explains the differences observed and how they can be modified to help men and women more fairly.

5.1.4 Organizational Challenges and Limitations

Despite their important contributions, environmental NGOs face significant challenges that limit their impact. Dependence on external funding, particularly from international donors, can create organizational vulnerabilities and shape their programmatic agendas.

Financial sustainability is one of the main challenges for environmental NGOs, especially those operating in developing countries, where resources for environmental programs are limited. This situation can result in the discontinuation of educational programs, the loss of trained personnel, and difficulties in maintaining long-term commitment to the beneficiary communities.

Hurtado Loaiza (2024) examines transformative environmental education within a rural Colombian context, identifying several challenges. The study reveals that environmental education is often disconnected from local contexts and the social realities of school and community environments, highlighting deficiencies in curriculum content and limited outreach efforts.

5.1.5 Impact and Effectiveness Assessment

Measuring the impact of educational programs developed by environmental NGOs presents complex methodological challenges, particularly when assessing long-term changes in attitudes, behavior, and environmental practices. Many organizations have developed monitoring and evaluation systems that combine quantitative and qualitative indicators to capture both the immediate results and transformative impacts of their educational interventions.

Chisika and Yeom (2024) present an environmental education strategy for sustainable forest management in Kenya, which serves as an exemplar of systematic evaluation. Their study identified the key ingredients of effective forest and environmental education as formal education infrastructure, community participation panels, digital platforms, corporate responsibility initiatives, media, arts, policy advocacy, and research with feedback loops on initiatives.

The documentation and systematization of experiences have emerged as important practices among environmental NGOs, enabling the construction of organizational knowledge and the transfer of learning between different implementation contexts. This practice has contributed to the development of conceptual and methodological frameworks that have influenced the field of environmental education.

5.2 COMMUNITY MOVEMENTS AND LOCAL LEADERSHIP

Community-based movements have established themselves as central expressions of social action addressing contextual environmental issues while serving as fertile ground for developing environmental education processes rooted in the particularities of each territory (Chawla, 2020). Their relevance lies in their ability to articulate citizen participation in the construction of localized responses to environmental challenges that directly affect local populations.

These initiatives generally emerge as collective responses to specific environmental conflicts or deterioration that impact the daily lives of the communities. Over time, they have become spaces for shared learning that integrate technical knowledge, local wisdom, and lived experiences. In this way, community movements transcend protest or corrective action by promoting deeply contextualized educational processes that strengthen the understanding of the environment and the collective capacity to intervene in an informed and sustainable manner.

5.2.1 Distinctive Features of Environmental Community Movements

A distinctive feature of community movements focused on environmental care is their internal origins within communities, their close connection to the territory, and their ability to mobilize and coordinate local resources around shared goals of environmental protection and management. These initiatives often arise when the population perceives environmental threats that compromise their quality of life or, conversely, when they identify concrete opportunities to improve the ecological conditions of the space they inhabit.

Similarly, these movements are characterized by their ability to integrate diverse knowledge sources into educational practices. Their dynamics involve the interaction between traditional knowledge, based on the historical and everyday experiences of the territory, and the technical and scientific contributions of external sources. This integration fosters bidirectional learning processes in which the community acts not only as a recipient of information but also as an active participant.

In the study by Kurpayanidi et al. (2024), digital transformation in environmental education highlights this complexity: While the incorporation of digital technologies can expand learning opportunities and training for new professionals, it also carries risks, including the possible loss of social interaction and a decline in the quality of educational processes, if not implemented critically and contextually.

5.2.2 Local Leadership and Sustainability of Community Education Processes

The role of local leadership is critical to ensuring that environmental education ideas emerging from neighborhood groups are sustained, connected, and expanded. Those who assume these roles are important because they bring diverse people together, lead group projects, and leverage diverse resources to achieve their goals. In this way, they help create a supportive environment in which learning matters and endures.

However, environmental leadership within a community does not emerge quickly; it is gradually built from the center. Its strength lies in its deep knowledge of the area, commitment to the common good, and ability to foster collaboration. In this way, those who lead understand how to speak well, work together, solve problems, and prepare themselves to communicate with others. With this in mind, Sandha et al. (2024) conducted a study on incorporating indigenous peoples' traditional knowledge of nature into environmental education in India, thereby demonstrating the importance of community leaders in the learning process.

The authors argue that this knowledge demonstrates a deep understanding of how to interact with nature, including its possibilities, limits, and the dangers of harming it. Simultaneously, they emphasize that staying alive requires not only respecting their cultural preferences but also holding official roles that protect them and ensure their transmission to future generations. Taken together, these ideas reinforce the notion that good environmental education must combine local knowledge with fixed rules that ensure continuity and validity over time.

5.2.3 Participatory Methodologies and Collective Knowledge Production

Neighborhood groups promote collaborative learning, in which knowledge is collectively built and sustained through democratic dialogue. In practice, this often takes the form of activities such as social mapping, analysis of neighborhood dynamics (how it is organized, which networks exist, and what tensions run through it), sharing experiences, and collective planning (Fraisl et al., 2022). This approach not only encourages people to take an active role in their learning process but also strengthens co-responsibility: The idea that environmental and social challenges are best addressed when commitments, tasks, and decisions are shared.

Within this framework, collaborative mapping has established itself as a particularly valuable educational strategy because it allows residents to represent their territory from their own perspective and, from there, to discuss and understand the connections between the environment, social fabric, and local economy. By mapping problems and resources, such as pollution hotspots, green areas, unsafe zones, mobility routes, meeting places, or available services, situated diagnoses are generated that facilitate the prioritization of real needs. In addition, these activities help highlight both the risks and opportunities of the environment, identify key actors, and agree on viable lines of action. Consequently, community groups not only identify relevant issues but also develop the capacity to design, implement, and monitor environmental initiatives tailored to their local conditions, thereby gaining greater social legitimacy and a greater potential for continuity over time.

The effectiveness of these ideas is supported by Imran et al. (2024), who examined how environmental education transforms the way people value and approach caring for the planet. Their data show that environmental education helps prepare young people for the future and that there is a connection between the ideas used in schools and those teachers consider when encouraging students to take care of the planet.

5.2.4 Articulation between Traditional Knowledge and Environmental Education

Community initiatives have played a fundamental role in identifying, recovering, and valuing traditional knowledge in environmental education processes. These experiences demonstrate that inherited knowledge of natural resource use, ecological cycles, and local environmental stewardship practices is essential for strengthening and diversifying current educational proposals.

The integration of this type of knowledge into community environmental education programs has led to the development of pedagogical approaches with a cross-cultural perspective, which recognize that environmental understanding is shaped by diverse cultural frameworks and that effective education must honor and bridge these different ways of knowing. These intercultural approaches have proven particularly valuable in contexts where indigenous and local communities maintain rich ecological knowledge systems that complement and, in some cases, challenge conventional scientific perspectives. By creating spaces for dialogue between these knowledge traditions, community environmental education programs foster mutual learning processes that strengthen both the cultural identity of participants and their capacity to address environmental challenges through locally appropriate strategies.

Mendoza et al. (2024) demonstrate the integration of environmental education with meaningful life experiences that promote environmental action among Filipinos. Their research indicates that while formal education is effective in enhancing technical skills, nonformal and informal educational approaches play a crucial role at various stages of an individual's life in shaping environmental actions.

5.2.5 Challenges to the Sustainability of Community Movements

The sustainability of community environmental movements faces multiple challenges, including resource availability, leadership continuity, coordination with public policies, and the ability to adapt to changes in the local dynamics (Redman et al., 2018). These challenges require specific strategies that strengthen the organizational capacities of communities and ensure the permanence of environmental education processes.

Training new leaders and the intergenerational transfer of environmental knowledge and commitments pose particular challenges for these movements. Communities have developed strategies, such as establishing community schools for environmental leadership, mentoring programs, and spaces for youth participation to ensure the continuity of organizational and educational processes.

Kerdsap et al. (2023) investigate strategies for mitigating the ecological footprint of a university farm through environmental education. Their research indicates that post-activity, students exhibited an increased understanding of ecological footprints, and the level of student engagement was notably high.

5.3 MEDIA AS EDUCATIONAL AGENTS

The media have evolved into key players in the environmental education ecosystem, fulfilling roles that extend beyond simply transmitting information to become true educational agents that influence the formation of environmental perceptions, attitudes, and behaviors of the public (Rousell & Cutter-Mackenzie-Knowles, 2020). Their ability to reach massive and diverse audiences positions them as powerful tools for democratizing environmental knowledge and building collective awareness of contemporary challenges.

5.3.1 Transformation of the Environmental Communication Paradigm

Environmental communication has progressively shifted from conventional models based on the linear dissemination of messages to more dialogical and inclusive perspectives, in which audiences no longer passively receive messages but actively construct meanings associated with the environment (Wamsler et al., 2021).

This change in mindset stems from something we see more clearly: Providing data is almost never sufficient to change people's behavior in the long term. Therefore, communication methods that encourage deeper reflection and connect individuals' thoughts with their feelings, culture, and lived experiences are becoming increasingly common.

From this perspective, it is understood that how people perceive and interpret environmental messages is strongly influenced by their social lives and daily experiences. Therefore, the most effective way to communicate clearly is to encourage people to participate, debate, and think critically. This is not a bonus but a key component in truly changing our impact on the planet. Similarly, many media outlets have begun using narrative techniques to make environmental issues easier to understand and more relatable, connecting major global issues to local experiences and helping audiences relate environmental topics to their daily lives.

The need for this transformation in environmental communication is further underscored by research on how environmental knowledge is currently transmitted through formal channels. Piscová et al. (2023), in their examination of environmental education in Slovak primary schools, found that environmental education remains poorly defined and often depends on individual teachers' preparation and motivation rather than on systematic communication strategies. Their study reported that students learn about their environment in a fragmented manner across isolated subjects—primarily in natural sciences and geography, and to a lesser extent in chemistry—rather than through integrated, narrative-driven approaches. These findings from formal education contexts reinforce the argument that media organizations must adopt more comprehensive, unified, and narrative-based communication strategies to address the fragmentation that characterizes conventional environmental education.

5.3.2 Diversification of Platforms and Audiences

The digital transition has diversified how we talk and teach about the environment, expanding access for people who were previously unaware of the traditional news. Simultaneously, online spaces have encouraged the creation of groups to learn about the environment. In these groups, people can share their experiences and resources and organize joint activities to care for the planet. In the university setting, the report by Bayaer and Cui (2023) is a good example of this; their results show that environmental education among university students positively affects the creation of green spaces. In addition, factors such as student satisfaction and learning enhance this relationship, contributing to the real environmental impact of education and influencing pedagogical strategies, which tend to change with the environment.

5.3.3 Media Teaching Strategies

The media have implemented educational strategies that leverage the unique characteristics of each channel and format to broaden their reach and strengthen the impact of environmental education content. This perspective stems from the understanding that each digital platform offers unique ways to interact with and learn about the environment, requiring teaching methods that align with its presentation style, pace, and usage patterns. Within this framework, ideas that incorporate game elements into discussions about the planet stand out, seeking to encourage curiosity, action, and participation from viewers through what appear to be games.

Similarly, the integration of deep vision devices has enabled immersive experiences that place people in natural environments, promoting a more practical understanding of current environmental issues. In addition, these narrative methods have created learning networks distributed across various interconnected platforms, reinforcing ideas and encouraging greater public participation in spaces where these issues are debated. An example of this is the work of Waluya et al. (2023) on the convergence of culture and environmental education, which showcases these teaching approaches. Their analysis of the Baduy indigenous people concluded that the skills and talents created by culture in terms of human adaptation and environmental education can be effectively transmitted through various media.

5.3.4 Collaboration with Educational Institutions

In the field of environmental stewardship, a new idea is gaining traction: Forging significant agreements between the media (especially the press) and educational institutions to promote environmental education. The rationale behind this collaboration is simple: Schools and universities provide technical, pedagogical, and practical knowledge for education, whereas the media offers a platform for dissemination, storytelling, and reaching a wide audience. By working together, both sectors can create and distribute enhanced educational materials (such as news articles with educational ideas, fact sheets, informative illustrations, activity guides, and interactive content) that can influence not only students but also families and other groups.

Simultaneously, many media outlets are developing projects and online resources designed to support environmental stewardship through education. These initiatives often include explanatory videos, downloadable classroom materials, digital files, and websites with activities, all of which contribute to improving teaching and learning. Furthermore, this approach makes it easier to combine in-person and remote learning: Students can access content with videos and audio at home or in a research space and then discuss, study, or use it in class with practical activities. This blended learning approach often leads to a better understanding of the natural world because it fosters more continuous engagement with environmental issues and provides more opportunities to observe, reflect, and participate.

Nedurmagomedov and Ataev (2023) explore these ideas further from their perspective, demonstrating how collaboration between education and media can be a viable path to creating a better future for the planet. The authors argue that environmental education should look to the future: Teaching with an eye to what is coming, but without neglecting the preservation of nature and the important past events that have occurred in the region. From this perspective, teamwork between entities not only seeks to "give information" but also seeks to change what is done and the values, helping to create more planet-centered classes using active teaching methods and media that make learning go beyond the classroom.

5.3.5 Ethical and Informational Quality Challenges

The expansion of news channels as key players in environmental education has highlighted serious challenges in terms of accuracy and a solid scientific basis for their content. In this context, news channels face the task of translating complex environmental issues into easily understandable and engaging messages without resorting to shortcuts that divert attention or diminish the importance of the issue. Achieving this balance requires closer collaboration with scientists, educators, and storytellers to ensure that the information presented maintains theoretical rigor while remaining accessible and capable of educating and influencing society positively.

Montenegro dos Reis et al. (2022) illustrated these difficulties in their study on teaching standards, management, and laws. Their analysis of ideas of totality, variety, and democracy in national environmental education standards found that while environmental knowledge is correctly named in teaching standards, its implementation remains inconsistent.

5.3.6 Measuring the Educational Impact of Media

Calculating the impact of environmental content in the media on learning presents significant challenges, particularly methodological ones, when assessing whether changes in knowledge, attitudes, and behaviors stem solely from viewing these programs or films. Current reporting methods are developing ways to assess this impact by combining the reach of information with data that delves into the educational and behavioral effects of what is presented.

Yekimov et al. (2022) used project-based learning to improve technical students' environmental education, underscoring the importance of this assessment approach. Their analysis concluded that teaching environmental topics to technical students should consider methods relevant to their professional futures, and that incorporating environmental issue projects into their learning provides students with the opportunity to build on their existing knowledge.

5.4 PRIVATE ENTERPRISE AND THE EVOLUTION OF ENVIRONMENTAL SOCIAL RESPONSIBILITY

Today, the business world plays a key role in environmental education. This is due to laws and the demands of customers and partners, who want companies to adopt methods that protect the planet (Lozano et al., 2019). This change has made environmental stewardship a central concern for business organizations, enabling new options and supporting comprehensive plans for environmental education.

Social and environmental issues in business management have changed significantly over time. They are viewed and implemented differently, moving from isolated actions to systems that prioritize the common good of society. In this context, green and social goals are integrated into business goals. The idea is that sustainability is not only something we must do but also key to competing and making money in the future. With this understanding, companies have a clearer perspective: Their production and sales activities impact on the environment negatively.

These impacts must be systematically managed. This awareness has driven initiatives to mitigate environmental damage, and environmental education is a key tool in these plans, helping reduce negative impacts. It also empowers people, both inside and outside the company. This helps build strong relationships with people and other groups. Therefore, new educational ideas have emerged in this field. They go beyond merely repairing the damage. They seek to provide robust and active support for sustainable development.

Ayres et al. (2022) examine the case of environmental education in São Tomé and Príncipe, highlighting that nonformal environmental education initiatives commenced in the 1990s. This development coincided with the initial efforts to engage with nature conservation through scientific approaches. They found that these efforts have grown considerably over the last ten years. This aligns with the need to include and encourage the local population. This is how better care is achieved, and conservation plans are guaranteed to be successful.

5.4.1 Modalities of Corporate Participation in Environmental Education

Companies have developed multiple forms of participation in environmental education, including funding educational programs, directly implementing educational initiatives, creating partnerships with specialized organizations, and integrating educational components into their value chains. Each form has specific characteristics in terms of scope, impact, and sustainability.

Corporate funding of environmental education programs developed by NGOs, academic institutions, and community organizations is a common form of corporate engagement. This approach enables companies to contribute to environmental education by leveraging the expertise and credibility of specialized organizations while developing their internal capacity for social and environmental program management.

Carvalho et al. (2021) present an environmental education and communication initiative focused on migratory fish and fishing communities, illustrating these modalities. Their research identified a significant enhancement in ocean and estuarine literacy and highlighted the efficacy of social media as a communication tool, particularly within the scientific community.

5.4.2 Internal Corporate Environmental Education Programs

An important dimension of corporate engagement in environmental education is developing internal programs targeting employees, suppliers, and business partners. These programs seek to develop environmental competencies within organizations, promote environmentally responsible behavior, and create corporate cultures aligned with sustainability principles.

Corporate environmental training programs have evolved from approaches focused on regulatory compliance to more comprehensive strategies that integrate the technical, ethical, and strategic dimensions of environmental management. These programs include training in clean technologies, resource management, environmental impact assessments, and sustainable product development.

Ardoin et al. (2020), in their systematic review of environmental education outcomes for conservation, provide evidence that comprehensive educational programs integrating multiple dimensions—cognitive knowledge, attitudinal change, and behavioral skills—consistently produce stronger outcomes than approaches focused on a single dimension. Their findings suggest that corporate environmental training programs are most effective when they move beyond regulatory compliance to address the full spectrum of environmental competencies, including technical knowledge, ethical reasoning, and practical skills for sustainable action. This evidence supports the evolution of corporate training from narrow compliance-focused approaches to more holistic strategies that develop genuine environmental commitment among employees, suppliers, and business partners.

5.4.3 Articulation with Formal Education Systems

Companies have developed strategies to link their environmental and social responsibility programs with formal education systems, creating opportunities for students at different educational levels to access practical and contextualized environmental learning experiences. These links include environmental internship programs, educational visits to corporate facilities, support for student research projects, and the development of specialized educational materials for environmental education.

Collaboration between companies and educational institutions has facilitated the development of technical environmental education programs that prepare students to work in economic sectors related to sustainability. These programs meet companies' needs for specialized human capital while providing students with training and employment opportunities.

Lozano et al. (2019) provide evidence supporting this articulation in their study of sustainability teaching in European higher education institutions. Their analysis demonstrated that effective connections between educational institutions and professional sectors require the systematic integration of sustainability competences into curricula through pedagogical approaches that combine theoretical foundations with practical, real-world applications. The authors found that institutions with stronger links to industry and community partners were more successful in developing students' capacity to address complex sustainability challenges, confirming that the articulation between formal education systems and external actors—including businesses—is essential for preparing professionals capable of leading sustainability transitions across diverse economic sectors.

5.4.4 Innovation in Educational Methodologies and Technologies

The industry has contributed significantly to advancing teaching strategies in environmental education, particularly by incorporating emerging technologies owing to its capabilities in research, development, and innovation (Petersen et al., 2020). This momentum has led to more agile educational materials, such as ecosystem and environmental process simulators, mobile applications for educational purposes, virtual reality experiences that enable the examination of complex ecological issues in immersive environments, and gamification approaches that increase motivation and support meaningful learning experiences.

Specifically, technology companies have strengthened these areas. They use new devices, including artificial intelligence (AI). They also rely heavily on large amounts of data and networks of connected devices. These tools have revitalized the learning process. They can be adapted to keep pace, tailored to each student's interests, and address individual weaknesses. Brundiers et al. (2021), in their agreed-upon reference framework for key competencies in sustainability education, emphasize that environmental education must be integrated across disciplinary boundaries rather than confined to specific subject areas. Their framework supports the principle that environmental themes should permeate all areas of instruction, serving as a cross-cutting dimension that connects scientific, technical, and humanistic knowledge within comprehensive educational standards.

This idea supports the value of comprehensive teaching methods, which new resources should support the latter. Overall, their work has expanded access to improved learning opportunities and provided teachers with additional options. This encourages the use of more modern approaches that incorporate student participation and are grounded in real-world evidence.

5.4.5 Challenges of Credibility and Greenwashing

Involving companies in environmental education generates revenue, new ideas, and increased visibility. However, this raises a sensitive issue: Public trust. Society views these ideas with caution. This is due to the danger of environmental deception in the real world (de Freitas Netto et al., 2020). Companies claim to be environmentally responsible, but their actions do not always align with their statements. If people see that what they say is not what they do, education loses its value. The educational effect is thus weakened.

Therefore, various groups are improving their account presentation and enhancing transparency. They seek to maintain this trust. Various methods have been used to achieve this. They commission external experts to review their work and provide feedback. They issue reports on their activities, which are then approved by experts. They also adhere to recognized environmental certification programs. In short, these methods seek to reduce the opacity of what is happening in the model. They seek to showcase their achievements. They want to demonstrate that their public statements reflect their daily lives. This fosters greater public confidence in the environmental education programs of companies.

Following this line of thinking, Golovanov et al. (2021) examined how to teach ecology to young children using various resources. This study highlights another challenge: Ensuring that teaching is sound and effectively implemented in practice. The results indicate that few people successfully perform ecological tasks with children. This reveals deficiencies in the planning of ideas and in the preparation of those who implement them in the field. This evidence confirms that for companies to contribute to ecological education, more is needed than simply wanting to do so and being seen doing it. Sound guidelines for good teaching are, therefore, essential. Those working in this field require adequate training. In addition, verifiable methods exist to ensure that everything is correct, useful, and produces lasting results.

5.4.6 Measuring Impact and Social Return on Investment

Companies have developed methods to measure the impact of environmental education (Stern et al., 2014). These methods aim to quantify the social and environmental effects of educational projects and assess their social benefits. These systems combine existing measures of school impact with established business metrics, such as brand image, stakeholder satisfaction, and the extent to which these initiatives contribute to the company's environmental goals.

Calculating the social benefits of environmental education is difficult. Measuring long-term environmental and social benefits is a complex process. Companies already use techniques to achieve this. They study social costs and gains, value the services provided by nature, and measure group effects to evaluate the effectiveness of environmental education programs.

Collado et al. (2020) demonstrated the importance of this. This study examined the impact of nature education by observing how it changed children's thinking and behavior. Their randomized group experiment yielded a key finding: Children demonstrated a better understanding of the environment in the group that learned through direct experience with nature. This level of understanding was higher than that in the group without such teaching.

5.5 CONCLUSION

The analysis of various civil society actors involved in environmental education reveals a diverse and constantly evolving network that substantially expands and complements the work of formal education systems. The plurality of approaches, resources, and capacities these actors contribute helps broaden the scope and depth of educational initiatives, thereby driving the development of more comprehensive proposals with greater transformative potential.

NGOs have established themselves as key drivers of pedagogical innovation and the incorporation of critical perspectives in environmental education. Their flexible structure and focus on social transformation allow them to design and implement programs that link learning to collective action and community change. However, their dependence on external funding and the fragility of their financial sustainability models highlight the need for institutional strengthening strategies that guarantee the continuity, operational stability, and sustained impact of these interventions.

Community movements and local leadership constitute the most deeply rooted forms of territorial expression in environmental education. These actors demonstrate that educational processes acquire greater transformative power when communities take a leading role in defining their priorities, building knowledge, and managing their territories. The convergence of traditional knowledge and technical expertise that characterizes these experiences gives rise to intercultural pedagogical approaches that enrich conventional perspectives on environmental education.

However, their consolidation faces challenges related to developing organizational capacity and effectively aligning with institutional frameworks and public policies to expand their reach.

Meanwhile, the media have evolved into more complex forms of action as educational agents, leveraging current technological resources to expand access to environmental education. Their capacity for mass dissemination and ability to adapt to diverse audiences make them strategic actors in constructing a shared environmental consciousness. Despite this, challenges remain in the quality, accuracy, and interpretation of information, as well as in strengthening environmental media literacy skills that enable audiences to critically evaluate the messages they receive.

The incorporation of the business sector into environmental education reflects an evolution toward more integrated and strategic corporate social responsibility models. Companies provide funding, technological infrastructure, and organizational capacity that can significantly enhance educational initiatives. However, the legitimacy of their participation depends on genuine commitment to transparency, consistency between discourse and practice, and accountability, which are essential elements in countering greenwashing and strengthening public trust.

Collaboration among these diverse actors presents both opportunities and challenges for environmental education. Establishing partnerships, collaborative networks, and mechanisms for cross-sector coordination can generate synergies that multiply the collective impact. This requires collaborative governance frameworks that recognize the diversity of approaches and capacities while facilitating convergence toward shared goals.

Furthermore, assessing the joint impact of these actors poses significant methodological challenges. It is necessary to promote evaluation frameworks that account for the specific outcomes of each type of actor and the emergent effects of their interactions. These frameworks must integrate quantitative and qualitative indicators to fully understand civil society's contributions to strengthening environmental education and achieving Sustainable Development Goals.

In summary, civil society actors are fundamental pillars of the environmental education ecosystem, contributing to methodological diversity, institutional flexibility, and innovative capacity that complement the strengths of formal education. Strengthening and strategically coordinating these efforts are essential for building more inclusive, effective, and transformative environmental education systems that can respond to current and future environmental challenges.

REFERENCES

Ardoin, N. M., Bowers, A. W., & Gaillard, E. (2020). Environmental education outcomes for conservation: A systematic review. *Biological Conservation*, *241*, 108224. https://doi.org/10.1016/j.biocon.2019.108224

Ayres, R., Aragão, J. C., Carvalho, M., Gouveia, F., Matilde, E., Panisi, M., Sacramento, J., & Schmitt, V. (2022). Environmental education in São Tomé and Príncipe: The challenges of owning a unique biodiversity. *Biodiversity of the Gulf of Guinea Oceanic Islands*, 671. https://doi.org/10.1007/978-3-031-06153-0_25#DOI

Bayaer, & Cui, D. (2023). The impact of college students' environmental education on green infrastructure. *Educational Administration: Theory and Practice*, *29*(3), 196–215. https://doi.org/10.52152/kuey.v29i3.654

Brundiers, K., Barth, M., Cebrián, G., Cohen, M., Diaz, L., Doucette-Remington, S., Dripps, W., Habron, G., Harré, N., Jarchow, M., Losch, K., Michel, J., Mochizuki, Y., Rieckmann, M., Parnell, R., Walker, P., & Zint, M. (2021). Key competencies in sustainability in higher education—toward an agreed-upon reference framework. *Sustainability Science*, *16*(1), 13–29. https://doi.org/10.1007/s11625-020-00838-2

Cândido Vendrasco, N., & Guerrero, G. (2025). Non-formal educators' roles in environmental education: Insights from Latin America. *British Educational Research Journal*, *n/a*(n/a). https://doi.org/10.1002/berj.70001

Carvalho, S. C., Braga, H. O., de Santa-Maria, S., Fonte, B., Pereira, M. J., García-Vinuesa, A., & Azeiteiro, U. M. (2021). An environmental education and communication project on migratory fishes and fishing communities. *Education Sciences*, *11*(7), 337. https://doi.org/10.3390/educsci11070337

Chawla, L. (2020). Childhood nature connection and constructive hope: A review of research on connecting with nature and coping with environmental loss. *People and Nature*, *2*(3), 619–642. https://doi.org/10.1002/pan3.10128

Chisika, S., & Yeom, C. (2024). Optimizing environmental education and awareness strategies for sustainable forest management in Kenya: Lessons from Cherangany, Mt. Kenya, Aberdares, and Kakamega Forest Ecosystems. *Visions for Sustainability*, *21*, 8830. https://doi.org/10.13135/2384-8677/8830

Collado, S., Rosa, C. D., & Corraliza, J. A. (2020). The effect of a nature-based environmental education program on children's environmental attitudes and behaviors: A randomized experiment with primary schools. *Sustainability*, *12*(17). https://doi.org/10.3390/su12176817

de Freitas Netto, S. V., Sobral, M. F. F., Ribeiro, A. R. B., & Soares, G. R. D. L. (2020). Concepts and forms of greenwashing: A systematic review. *Environmental Sciences Europe*, *32*(1), 19. https://doi.org/10.1186/s12302-020-0300-3

Fraisl, D., Hager, G., Bedessem, B., Gold, M., Hsing, P.-Y., Danielsen, F., Hitchcock, C. B., Hulbert, J. M., Piera, J., Spiers, H., Thiel, M., & Haklay, M. (2022). Citizen science in environmental and ecological sciences. *Nature Reviews Methods Primers*, *2*(1), 64. https://doi.org/10.1038/s43586-022-00144-4

Golovanov, V. P., Bykovskaya, T., Panteleeva, N., Fedorova, S. Y., & Tretyakov, A. (2021). Methodological strategy of modern preschool environmental education based on the resource approach: Prospects of science and education, 1. *Perspektivy Nauki i Obrazovania—Perspectives of Science and Education*, *10*, 299–317. https://doi.org/10.32744/PSE.2021.1.21

Hurtado Loaiza, C. V. (2024). Hacia una Educación Ambiental Transformadora en un Contexto Rural de Colombia. *European Public & Social Innovation Review*, *9*, 1–18. https://doi.org/10.31637/epsir-2024-1084

Imran, M., Almusharraf, N., & Abdellatif, M. S. (2024). Education for a sustainable future: The impact of environmental education on shaping sustainable values and attitudes among students. *International Journal of Engineering Pedagogy*, *14*(6), 155–171. https://doi.org/10.3991/ijep.v14i6.48659

Kerdsap, T., Tanpichai, P., & Yingyuad, N. (2023). Reducing the ecological footprint in a university farm through environmental education process. *Environment Asia*, *16*(3). https://doi.org/10.14456/ea.2023.32

Kurpayanidi, K., Khamdamova, S., Ergashev, D., Nishonov, U., Kholmatova, D., & Khoshimov, A. (2024). A balanced view of digital transformation in environmental education. E3S Web of Conferences.

Lozano, R., Barreiro-Gen, M., Lozano, F. J., & Sammalisto, K. (2019). Teaching sustainability in European higher education institutions: Assessing the connections between competences and pedagogical approaches. *Sustainability*, *11*(6), 1602. https://doi.org/10.3390/su11061602

Mendoza, I. A. F., Gotangco Gonzales, C. K., & Favis, A. M. T. (2024). Experience-based environmental education and significant life experiences that influence environmental action among Filipinos. *Journal of Environmental Studies and Sciences*. https://doi.org/10.1007/s13412-024-00956-8

Mogren, A., Gericke, N., & Scherp, H.-Å. (2019). Whole school approaches to education for sustainable development: A model that links to school improvement. *Environmental Education Research*, *25*(4), 508–531. https://doi.org/10.1080/13504622.2018.1455074

Montenegro dos Reis, R. F., Rodrigues dos Reis, M., & Peregrino Montenegro, P. (2022). Política, administração e direito educacional: noções de holismo, pluralidade e democracia na política nacional de educação ambiental. *Revista Brasileira de Políticas Públicas*, *12*(3). https://doi.org/10.5102/rbpp.v12i3.7904

Nedurmagomedov, G., & Ataev, Z. (2023). National and regional experience in the implementation of environmental education for sustainable development and preservation of regional natural and cultural heritage. *South of Russia: Ecology, Development*, *18*(1), 140–156. https://doi.org/10.18470/1992-1098-2023-1-140-156

Petersen, G. B., Klingenberg, S., Mayer, R. E., & Makransky, G. (2020). The virtual field trip: Investigating how to optimize immersive virtual learning in climate change education. *British Journal of Educational Technology*, *51*(6), 2099–2115. https://doi.org/10.1111/bjet.12991

Piscová, V., Lehotayová, J., & Hreško, J. (2023). Environmental education in the school system at elementary schools in Slovakia. *European Journal of Science and Mathematics Education*, *11*(4), 650–671. https://doi.org/10.30935/scimath/13377

Redman, E., Wiek, A., & Redman, A. (2018). Continuing professional development in sustainability education for K-12 teachers: Principles, programme, applications, outlook. *Journal of Education for Sustainable Development*, *12*(1), 59–80. https://doi.org/10.1177/2455133318777182

Repetto, M., Bianco Prevot, A., Brizio, A., Boldi, A., Talarico, M., Stanchi, S., Palma, D., & Acquadro Maran, D. (2024). Digital escape rooms: A resource for environmental education. *Sustainability*, *16*(19), 8525. https://doi.org/10.3390/su16198525

Rousell, D., & Cutter-Mackenzie-Knowles, A. (2020). A systematic review of climate change education: Giving children and young people a 'voice' and a 'hand' in redressing climate change. *Children's Geographies*, *18*(2), 191–208. https://doi.org/10.1080/14733285.2019.1614532

Sandha, H. S., Khare, P., & Sharma, P. (2024). Integrating Traditional Ecological Knowledge (TEK) with environmental education in India. In T. Kochetkova (Ed.), *Fostering an Ecological Shift Through Effective Environmental Education* (pp. 189–205). IGI Global Scientific Publishing. https://doi.org/10.4018/979-8-3693-2577-3.ch012

Stern, M. J., Powell, R. B., & Hill, D. (2014). Environmental education program evaluation in the new millennium: What do we measure and what have we learned? *Environmental Education Research*, *20*(5), 581–611. https://doi.org/10.1080/13504622.2013.838749

Wals, A. E. J., & Benavot, A. (2017). Can we meet the sustainability challenges? The role of education and lifelong learning. *European Journal of Education*, *52*(4), 404–413. https://doi.org/10.1111/ejed.12250

Waluya, B., Malihah, E., Ruhimat, M., & Wiyanarti, E. (2023). Cultural ecology and environmental education: Lesson learned from Baduy indigenous community. *Indonesian Journal of Geography*, *55*(1), 88–97. https://doi.org/10.22146/ijg.77203

Wamsler, C., Osberg, G., Osika, W., Herndersson, H., & Mundaca, L. (2021). Linking internal and external transformation for sustainability and climate action: Towards a new research and policy agenda. *Global Environmental Change*, *71*, 102373. https://doi.org/10.1016/j.gloenvcha.2021.102373

Wardhani, D., Arisanty, D., Nugroho, A., & Utami, U. (2024). Environmental education model based on local wisdom of the Dayak Paramasan tribe Indonesia. *Nature Environment and Pollution Technology, 23*(4), 2259–2272. https://doi.org/10.46488/NEPT.2024.v23i04.030

Yekimov, S., Aleksey, M., Natalia, G., Bruslova, O., & Bruslov, A. (2022, October 13). Using the project methodology to improve environmental education for students of technical specialties. IOP Conference Series: Earth and Environmental Science, Atlanta, GA, USA.

6 Institutional Framework and Public Policies

6.1 NATIONAL AND INTERNATIONAL ENVIRONMENTAL EDUCATION POLICIES

6.1.1 Evolution of the International Policy Framework

The evolution of international environmental education policies has been shaped by a series of global conferences and declarations that set principles, objectives, and guidelines for action by the international community (Kopnina, 2020). Research by Mendonça et al. (2025), in the context of geoethics and sustainability, to address challenges in environmental education and achieve the Sustainable Development Goals (SDGs), emphasizes that geoethics, which explores the ethical implications of Earth sciences, requires philosophical reflection to deepen our understanding of the field. The 1972 Stockholm Conference marked the formal beginning of global attention to environmental education as an indispensable element of governance.

The conference concluded that environmental education should target young people and adults, including less privileged segments of society, to shape public opinion and informed, responsible behavior that supports environmental protection. Lee and Kim (2025) reveal that "gaps persist in the implementation and effectiveness of the CNMS project, requiring further analysis informed by the complex interactions between actors and policy factors." This analysis highlights the importance of defining clear project objectives and a shared vision for environmental education and carbon neutrality in schools.

6.1.2 Contemporary Developments in International Policy

Contemporary progress in international environmental education policy has been significantly influenced by the 2030 Agenda for Sustainable Development and the Paris Agreement on Climate Change. These models have transformed environmental education into one that is intertwined with global sustainability and connects educational priorities to broader initiatives on climate action and sustainable development.

Research has consistently demonstrated that environmental education can effectively change specific environmental practices, including waste management behaviors. Ardoin et al. (2020) conducted a systematic review of environmental education outcomes for conservation and found that well-designed educational programs have a positive and significant impact on environmental knowledge, pro-environmental attitudes, and behavioral intentions. Their analysis further revealed that educational interventions are most effective when they integrate cognitive, affective, and behavioral components and that the strength of educational outcomes is moderated by

DOI: 10.1201/9781003779667-8

contextual factors such as the duration of the intervention, the pedagogical approach employed, and the cultural values of the target community. These findings demonstrate how international policy frameworks, such as the 2030 Agenda for Sustainable Development and the Paris Agreement, must be supported by evidence-based educational strategies that address both the cognitive and emotional dimensions of environmental behavior change.

UNESCO is the source of international environmental policy and communication, not least through the Education for Sustainable Development program and the Decade of Education for Sustainable Development (2005–2014) (Leal Filho et al., 2020). The role of UNESCO, continuing education, and the delivery of environmental education in Botswana provides a critical lens on these international initiatives, as mentioned by Velempini and Maruatona (2025).

Their analysis shows that, although the report has not been formally endorsed, Botswana has benefited directly and indirectly from UNESCO's technical assistance over time. However, it still faces political, practical, and technical challenges in delivering continuing and environmental education to teachers. This is further evidence of the possibilities and limitations of international frameworks in influencing national policies and practices.

6.1.3 Variations in National Implementation

The national implementation of environmental education policy at this level shows marked variation, reflecting differences in political, economic, social, and cultural contexts. These differences can facilitate a comparative understanding of the factors that enable or hinder the success of environmental education policies from truly succeeding.

Igei et al. (2024) studied the synergistic effects of nudges and prompts in environmental education and provided experimental evidence that different policy approaches exhibit varying efficacies. Their field experiment in Japan showed that students who received the prompt became significantly more environmentally conscious during the game and set more goals in the nudge goal-setting task.

The findings indicate that combining nudges with impetus sustains the long-term effects of environmental education on reducing plastic waste. Jaime et al. (2023) studied children's environmental education and pro-environmental behavior in Chile and demonstrated the potential of school environmental education programs to impact environmental attitudes and practices. Using a randomized field experiment, they found that the educational program had a substantial positive effect on children's knowledge, attitudes, and practices. Their multilevel analysis further revealed that the intensity and quality of program implementation at the school level moderated the magnitude of these effects, suggesting that national certification programs that establish clear standards for environmental education can serve as effective policy instruments for promoting pro-environmental behavior among children

6.1.4 Challenges in International Coordination

The international coordination of environmental education policies faces various challenges, including differences in national priorities, institutional capacities, and regulatory frameworks (Menton et al., 2020). These obstacles require global governance mechanisms that promote cooperation while respecting national sovereignty and the diversity of contexts.

Hurtado Loaiza (2024) highlighted some of these challenges. They found that "environmental education has been disconnected from local contexts, distancing itself from social reality in school and community environments, and showing gaps in curriculum content and limited community outreach." This finding highlights the need to create policies that integrate international frameworks with the specific realities of each community in Brazil.

6.1.5 International Monitoring and Evaluation Mechanisms

Challenges in building effective monitoring and evaluation processes for international environmental education interventions include the comparability of indicators across locations, data availability, and institutional capacity for data collection and interpretation.

Effective monitoring and evaluation of international environmental education interventions face significant challenges, including the comparability of indicators across diverse national contexts, data availability, and institutional capacity for systematic data collection and interpretation. Rieckmann and Barth (2022) emphasize that current monitoring frameworks often fail to capture the transformative dimensions of environmental education, focusing instead on quantitative participation metrics rather than meaningful learning outcomes. Their analysis reveals that developing robust evaluation systems requires international coordination to establish shared indicators while maintaining sufficient flexibility to accommodate regional differences in educational priorities and institutional capacities.

6.2 REGULATORY AND POLICY FRAMEWORK

6.2.1 Development of National Legal Frameworks

National legal frameworks for environmental education have evolved from voluntary approaches to more organized systems with specific obligations for educational institutions, government agencies, and other relevant actors (Lozano et al., 2019). This transformation illustrates the growing awareness that environmental education needs more than isolated self-regulated programs.

The research by Reis et al. (2024) has examined the complexities of integrated risk management, environmental education, and fuzzy logic. They state that fuzzy logic is incorporated into the conclusions section and the study proposal as a tool for designing and implementing risk management, considering the inherent complexity called the "risk society."

This perspective holds that, owing to their complexity, environmental problems require regulatory frameworks capable of operating under uncertainty and in situations that are not always fully defined. In this sense, the most robust legal frameworks typically include several components that address different dimensions of environmental education and are designed with the specific needs of all parties involved.

These elements may include guidelines for incorporating environmental education into formal education curricula, requirements for professional training programs, provisions for public funding, and mechanisms to facilitate coordination between government institutions and civil society organizations.

Kiraz and Danju (2025) examined the integration of digital education into environmental education, demonstrating how regulatory frameworks must adapt to technological advances. Their study also explored the media's role as an educational agent in fostering individual and social environmental awareness.

Finally, the study concludes that modern regulatory structures should account for both the opportunities and risks associated with digital technologies in environmental education.

6.2.2 Enforcement Mechanisms and Integration with Sectoral Regulatory Frameworks

This research identifies a key challenge in developing effective compliance and enforcement mechanisms for environmental education regulatory frameworks at both the state and national levels: Measuring educational outcomes is challenging. They may not materialize until long after the implementation. These mechanisms are intended to balance accountability and flexibility for innovation and local adaptation (Ardoin et al., 2018).

Yadav (2023) asserts that education is crucial for developing sustainable behavior and critical environmental awareness and is a primary instrument for addressing these issues. It can foster critical thinking and alter human behavior in the natural world.

This view emphasizes that compliance mechanisms must focus on transformative outcomes rather than procedural compliance to be effective. Many leading methods for addressing compliance use a combination of positive incentives, technical assistance, and collaborative monitoring to encourage continuous improvement rather than merely noncooperation with enforcement rules. These approaches recognize that institutional capacities are necessary before environmental education can be successfully implemented, which takes time.

Integrating environmental education regulations with broader sectoral legal frameworks, such as education in the environment, health, agriculture, and urban development, is the main challenge influencing the overall effectiveness of environmental education (Florencio da Silva et al., 2023).

6.2.3 Public Participation in Regulatory Development

A more collaborative approach to establishing regulatory frameworks for environmental education that encourages public participation has been suggested (Wamsler et al., 2021). This can help improve the legitimacy and effectiveness of environmental policies by integrating diverse viewpoints and increasing popular support for their adoption.

However, for public participation to be truly effective, it must be approached as an intentional and well-structured process. Ma et al. (2023) analyzed the moderating effect of environmental education on the maintenance of a broad environment, providing clear evidence on the relevance of involving citizens. Their findings show that environmental education significantly strengthens environmental knowledge and, especially among young people, promotes the adoption of more environmentally friendly lifestyles.

The findings also reported that pro-environmental intentions toward sustainable consumption and the impact of these intentions on sustainable consumption practices were greater among those with environmental education than among those without it.

6.2.4 Challenges in Multilevel Implementation

The relevance of such regulatory regimes for environmental education in multilevel governance settings presents unique challenges, namely, in the authority structure, the organization of authorities, the coordination of management across levels of government, and adaptation to different local contexts (Brundiers et al., 2021).

Obstacles must be overcome for such regulatory devices, which must protect national-level consistency while remaining flexible at the local level. Jaime et al. (2023) examined the extent to which school environmental education programs can increase children's and parents' pro-environmental attitudes, a multilevel challenge. In Chile, a randomized field experiment found that the educational program had a substantial positive effect on children's knowledge, attitudes, and practices but no effect on parental behavior. This suggests that regulatory approaches must address diverse audiences and contexts simultaneously.

6.3 FUNDING AND RESOURCES FOR EDUCATIONAL PROGRAMS

6.3.1 Public Funding Models

Public funding for environmental education is often provided through budgetary mechanisms, such as direct allocations to specific areas, competitive and public–private incentives, and mechanisms to internationalize funding for this purpose (Sonetti et al., 2020). The success of these models is largely determined by their ability to create incentives for innovation, accountability, and sustainability of practices in the future.

Ugolini and Pearlmutter (2022) explored new financing options for environmental education solutions by examining how location-based technologies can analyze urban threats and identify nature-based solutions. Their analysis of European educational projects shows that when delivered in an immersive and interactive manner, environmental education can inspire students to understand the ecological challenges we face and address them creatively through nature-based solutions.

Effective funding models typically incorporate measures for ongoing multiyear programs, allowing institutions to invest, build capacity, and evaluate long-term impacts. Matching or co-funding programs could also be included in such models, as they can motivate local engagement and diversify resource availability. Guerra et al.

(2022) studied these developments and highlighted the need for financing systems that can adapt to new crises while operating sustainably to support long-term goals.

6.3.2 Innovative Financing Instruments

The design of novel funding models for environmental education stems from the understanding that conventional government funding may be insufficient to meet the scale of environmental-education requirements. These include green bonds, payments for ecosystem services, environmental investment funds, and results-based payment methods.

Bushkova-Shiklina and Musikhina (2020) link environmental education for students as a means of increasing social capital, showing links between educational expenditures and broader social outcomes. Their analysis found a strong willingness among students to engage in selective waste collection alongside other environmental practices and a strong cognitive interest in this issue.

These findings suggest that novel payment forms should consider both direct educational benefits and positive social effects. Social impact funds are a promising approach to aligning private investment goals with measurable outcomes in environmental education. Such funds almost always require rigorous impact measurements and can incentivize innovation in teaching methods and assessment systems.

6.3.3 Challenges in Financial Sustainability

Long-term financial sustainability remains a persistent challenge for environmental training projects, particularly for initiatives implemented by social entities that may depend on unstable, external funding. These challenges require plans to reallocate resources and improve skills to generate income.

Bravo et al. (2020) analyzed the work of a network united in an environmental education project in southern Chile. This study found that the network fulfills the role of maintaining adherence to these four points by providing administrative assistance, leadership, skills development, and social support. This suggests that networks can be an important means of improving financial sustainability by sharing resources and skills.

6.3.4 Return on Investment Assessment

Assessing the return on investment for environmental training projects poses unique challenges, as many benefits are long-term, complex to measure, or manifest as positive externalities that benefit the general population rather than specific investors. These problems require the development of valuation models that capture social value alongside traditional economic benefits.

Assessing the return on investment for environmental education programs requires consideration of extended benefits that accumulate over time. Redman et al. (2018) examined the long-term impacts of sustainability education on preservice teachers and found that students who received systematic environmental training during their university education demonstrated progressively stronger sustainability competencies as they advanced through their programs. These findings suggest that investing in teacher training creates multiplier effects, as each trained educator can influence hundreds of students throughout their professional career, generating cumulative returns that far exceed the initial investment. These data suggest that investing in teacher training can create extended benefits throughout the working lives of these professionals.

6.3.5 Public–Private Partnerships

Public–private partnerships for financing environmental education have emerged as a key means of leveraging private capital while retaining state control over educational goals (de Freitas Netto et al., 2020). Such partnerships can provide access to technical expertise, delivery systems, and change capabilities from the private sector while ensuring that educational projects remain focused on the goals of general interest.

Araujo et al. (2020) studied the Paris Agreement through environmental education and the roles of different sectors in implementing global environmental policies. Their study argues that effective partnerships require clear governance structures that align private incentives with public objectives, while maintaining transparency and accountability.

6.3.6 Horizontal Coordination Mechanisms

Working together across government departments is very difficult for environmental education plans to be successful. This is because these plans almost always require subject-matter expertise and funding from multiple ministries and agencies. For this to work, each agency must be prevented from pursuing its own interests and encouraged to help one another.

International experiences in horizontal coordination offer valuable lessons for environmental education governance. Cebrián et al. (2021), in their analysis of current practices and future pathways toward competencies in education for sustainable development, found that effective institutional coordination requires the development of shared competency frameworks that facilitate communication and alignment across organizational boundaries. Their study identified key competencies that practitioners need to collaborate effectively across institutional contexts, suggesting that horizontal coordination depends not only on formal governance structures but also on the development of shared professional competencies among educators working in different institutional settings.

The most effective coordination mechanisms typically include formal governance structures, such as committees across ministries, expert groups, and mechanisms for sharing information, facilitating communication, and providing mutual support. These rules must be supported by clear directives, sufficient funding, and accountability for what we aim to achieve.

6.3.7 Multilevel Vertical Coordination

Collaboration between different levels of government, whether national, regional, or local, faces particular challenges in terms of environmental education due to the different authorities, capacities, and priorities that may exist at each level of government (Veidemane, 2022). For vertical coordination to function effectively, it is essential to strike a balance between national coherence and local adaptability while ensuring that all levels have the necessary resources and skills to perform their tasks.

Almeida (2017) explored policymakers' and practitioners' perceptions of the impact, evidence, and support for educators implementing environmental education for sustainability in India and offered valuable insights into these challenges. Their research focused on the conditions and elements that drive the development and implementation of environmental education policies.

The results reveal the main facilitators and barriers currently affecting teacher training in India, lessons learned globally on implementing reforms in the country, and future needs.

6.3.8 Participation of Nongovernmental Actors

The effective involvement of nongovernmental actors in cross-sectoral cooperation requires institutional structures that can integrate different forms of expertise, organizational models, and accountability frameworks. Such participation can significantly enrich policymaking processes by offering practical knowledge, beneficiary perspectives, and implementation skills that complement government resources.

The study conducted by Valenti et al. (2017) on the exclusionary and transformative aspects of adult environmental education in two nature reserves in Brazil investigated collaboration between the staff of these protected areas and local communities. They applied a critical communicative methodology and included 25 individuals with diverse connections to puma conservation and environmental education. The results indicate that dialogic interaction is necessary to jointly define community needs and find solutions through regional and international coordination.

Regional and international collaboration in environmental education has become increasingly relevant, given the transnational nature of many environmental challenges and the opportunities to exchange resources, expertise, and best practices among nations. This collaboration can be carried out through regional entities, professional networks, exchange programs, and collaborative research projects.

Regional and international collaboration has become increasingly relevant given the transnational nature of many environmental challenges and the opportunities to exchange resources, expertise, and best practices among nations. Menton et al. (2020), in their analysis of environmental justice and the SDGs, demonstrated that international collaboration in environmental education must account for cultural differences in environmental values, consumption patterns, and governance traditions. Their findings suggest that effective international agreements and collaborative frameworks must respect these differences while establishing shared principles that can guide coordinated action across diverse national and cultural contexts.

6.3.9 Integrated Information and Monitoring Systems

The development of integrated information and monitoring systems is key to effective coordination, as these tools facilitate data exchange, organize tasks, and enable the tracking of progress toward shared goals. Veidemane (2022) emphasizes that the development of internationally comparable indicators for education for sustainable development requires robust information systems capable of integrating diverse data types from multiple institutional sources while maintaining methodological rigor and cross-national comparability. These systems must be capable of handling diverse data categories, users with varying levels of technical skills, and privacy and security requirements.

6.3.10 Conflict Management and Dispute Resolution

Proper management of disagreements and problem-solving is an often underestimated yet fundamental aspect of cooperation between sectors, as participants may hold conflicting mindsets, approaches, and value systems. Effective methods for handling disputes can turn friction into opportunities for learning and creativity.

Junior et al. (2016) investigated the political dimensions of environmental education processes. Their research suggests that fostering public awareness from the early stages of education helps children develop political agencies, enabling them to question and influence their social environments. From this perspective, effectively resolving conflicts entails recognizing that environmental education is inherently political. Furthermore, because there are multiple ways to understand sustainability, it is essential to encourage dialogue and foster compromises among different positions.

6.4 CONCLUSION

6.4.1 Summary of Institutional Challenges

The analysis of the institutional context and public policies related to environmental education reflects a complex picture: Significant progress has been made, but challenges remain that require ongoing attention and innovative responses to address them (Chawla, 2020). Although international frameworks have provided guidance and supported national policies, their effective implementation remains difficult owing to limitations in institutional capacity, insufficient funding, and the need for better coordination among actors and across levels of government.

Cocks and Simpson (2015), in their philosophical foundations, emphasize that institutional frameworks should not focus solely on technical or administrative aspects but also on values and worldviews that, in practice, can directly influence policy implementation.

At the institutional level, specific challenges include strengthening horizontal coordination across government departments, developing more stable financing mechanisms to support long-term efforts, and designing regulatory frameworks that balance national coherence and local flexibility. These difficulties tend to be more pronounced in developing countries, where institutional capacity is weaker and resources are limited.

6.4.2 Critical Factors for Institutional Success

The research presented in this study highlights several key elements that influence the success of institutional structures for environmental education. First, strong political leadership with a clear vision is crucial for overcoming resistance within organizations and mobilizing the resources needed for efficient implementation. This type of leadership must operate across multiple levels of the political system and maintain its commitment over time, even across cycles.

Sustained political leadership is also reflected in how environmental education outcomes are perceived by different stakeholders. Ardoin et al. (2020) conducted a systematic review examining how participants and practitioners perceive the outcomes of environmental education programs. Their analysis revealed that while both groups consistently reported increased environmental awareness as the most frequently cited outcome, significant differences emerged in how each group valued behavioral change, community engagement, and long-term attitudinal shifts. These discrepancies suggest that effective leadership must facilitate ongoing dialogue between program designers and beneficiaries to ensure that institutional goals align with community needs and expectations.

However, the differences between their perspectives suggest a need for greater dialogue between professionals and participants. This reinforces the idea that effective leadership must facilitate open and honest conversations among the various actors involved. In addition, it is essential to strengthen institutional capacities to design, implement, and evaluate environmental education programs so that policies are translated into concrete actions. These capacities include competencies in environmental pedagogy, program management, evaluation, and promotion of participatory processes.

Finally, building broad alliances that integrate the government, civil society organizations, the private sector, and local communities can provide political continuity and resources to sustain long-term actions. To function effectively, these coalitions must be grounded in a shared vision of sustainability.

6.4.3 Directions for Future Institutional Development

The analysis identified several strategic lines that should be prioritized in future institutional strengthening to enhance the effectiveness of policy frameworks for environmental education. First, there is a significant opportunity to transform financing schemes, particularly by diversifying funding sources and designing incentives linked with verifiable outcomes. This type of innovation can significantly help consolidate the financial viability and sustainability of environmental education programs (Rousell & Cutter-Mackenzie-Knowles, 2020).

In this regard, the study conducted by Hyseni Spahiu et al. (2014) on environmental education in secondary schools in Kosovo provides key empirical evidence for understanding institutional development needs. The authors found that although the teachers selected topics that were highly relevant, contextualized, and linked to students' everyday experiences, fundamental content related to sustainability, such as biodiversity loss, was omitted.

It was also observed that certain basic elements, such as large class sizes and short activity durations, significantly reduced the likelihood of conducting outdoor activities, particularly those involving fieldwork. These data demonstrate that improving institutions requires consideration of both what is taught and the problems that hinder effective implementation of environmental education.

Furthermore, it emphasizes the importance of strengthening observation and evaluation methods, making them more robust and modern, so they can capture not only immediate learning but also how outcomes change over time in response to social change. The development of these methods is essential for better accountability and institutional learning; therefore, it is beneficial to use mixed-methods approaches that combine quantitative data with qualitative studies of change processes.

Finally, it is important to increase public participation in decisions regarding environmental education. If stakeholders are involved from the outset in planning objectives, methodologies, and rationale, the resulting policies can be more effective and gain broader acceptance. However, this participation must go beyond simply soliciting opinions; it must also include collaboration and support through programs so that people are recognized as an important part of the decision-making process.

6.4.4 Imperatives for Coordinated Action

The evidence presented in this section demonstrates the urgent need to promote collaborative initiatives that overcome current institutional divisions and foster genuine mutual support among entities, departments, and different levels of management. This unity stems from the understanding that effective environmental education is not based solely on educational policies but requires comprehensive transformations that address the root causes of environmental degradation and social inequality.

In this context, Hadjileontiadou et al. (2013) focused on improving accessibility for students with disabilities through information and communication technologies in environmental education, exemplifying the relevance of this comprehensive approach. Their analysis of MusicPaint software showed that, even though the number of participants in the experimental validation scenarios was small, the results demonstrated the high potential of this tool to expand access to environmental education experiences for students with disabilities. These findings reinforce the need for institutional frameworks to incorporate explicit provisions for equity and inclusion, which necessarily implies coordination among the education, environment, and disability sectors.

Similarly, the severity and urgency of current environmental problems, particularly climate change, accelerated biodiversity loss, and ecosystem degradation, require that these coordinated actions be implemented on an unprecedented scale and speed. However, existing evidence indicates that effective coordination cannot be established overnight. In contrast, establishing inter-institutional links, strengthening mutual trust, and harmonizing incentives and processes tend to progress gradually.

6.4.5 Vision for Transformative Institutional Frameworks

The emerging perspective on institutional frameworks for transformative environmental education underscores the need to move beyond traditional sectoral governance and adopt integrated approaches that address the systemic complexity of sustainability challenges. From this perspective, environmental education should not be conceived as a specialized and isolated field but rather as a cross-cutting component that dialogues and interacts with policies on economic development, urban planning, public health, and social justice (Whitburn et al., 2020).

The study by Fraijo-Sing et al. (2013), which focuses on the design of environmental education programs, provides relevant conceptual elements for understanding this vision of change. Based on a model of pro-environmental competence, the authors point out that effective programs must incorporate dimensions such as environmental knowledge, environmentally favorable attitudes, and the intention to act responsibly. However, they also highlight the importance of considering contextual factors that can enhance or limit the effective manifestation of such competence, reinforcing the need for comprehensive approaches that extend beyond the individual level.

This transformative conception also highlights the importance of democratizing governance processes in environmental education, creating meaningful opportunities for participation for historically marginalized communities, young people, and other groups with limited involvement in educational decision-making. The inclusion of these voices not only improves the relevance and effectiveness of policies but also aligns with the broader objectives of equity, social justice, and strengthening the community fabric.

Similarly, the proposed vision recognizes that institutional frameworks in this area must continuously adapt to new scientific advances, technological developments, and transformations in social and political contexts. This flexibility involves promoting organizational cultures oriented toward lifelong learning, informed experimentation, and critical reflection on educational practices.

The realization of this transformative approach requires sustained commitment from generations of policymakers, education professionals, and citizens willing to invest in long-term change. However, given the magnitude and urgency of current environmental problems, such investments are not only desirable but also indispensable to ensure the future well-being of both human societies and natural systems.

REFERENCES

Almeida, S. (2017). Policy-makers' and practitioners' perspectives on impact, evidence, and support for teacher educators implementing environmental education for sustainability in India. In J. Nuttall, A. Kostogriz, M. Jones, & J. Martin (Eds.), *Teacher Education Policy and Practice: Evidence of Impact, Impact of Evidence* (pp. 35–48). Singapore: Springer. https://doi.org/10.1007/978-981-10-4133-4_3

Araujo, L. M., Cordeiro Neto, M. D. R., & Seguin, E. (2020). A busca da efetividade do Acordo de Paris. *Revista Direito, Estado e Sociedade*, *0*(54). https://doi.org/10.17808/des.54.768

Ardoin, N. M., Bowers, A. W., & Gaillard, E. (2020). Environmental education outcomes for conservation: A systematic review. *Biological Conservation*, *241*, 108224. https://doi.org/10.1016/j.biocon.2019.108224

Ardoin, N. M., Bowers, A. W., Roth, N. W., & Holthuis, N. (2018). Environmental education and K-12 student outcomes: A review and analysis of research. *The Journal of Environmental Education*, *49*(1), 1–17. https://doi.org/10.1080/00958964.2017.1366155

Bravo, G. P., Andrade, R. R., & Hevia, N. B. (2020). Tejiendo raíces para la implementación: las funciones de una red colaborativa en un programa de educación ambiental al sur de Chile. *Revista Luna Azul*, (50), 174–196. https://doi.org/10.17151/luaz.2020.50.9

Brundiers, K., Barth, M., Cebrián, G., Cohen, M., Diaz, L., Doucette-Remington, S., Dripps, W., Habron, G., Harré, N., Jarchow, M., Losch, K., Michel, J., Mochizuki, Y., Rieckmann, M., Parnell, R., Walker, P., & Zint, M. (2021). Key competencies in sustainability in higher education—toward an agreed-upon reference framework. *Sustainability Science*, *16*(1), 13–29. https://doi.org/10.1007/s11625-020-00838-2

Bushkova-Shiklina, E., & Musikhina, T. (2020). Environmental education of students as a factor in social capital development (through the example of the practice of household waste separate collection). Perspektivy Nauki i Obrazovania. *Перспективы Науки и Образования Perspectives of Science and Education*, 289. https://doi.org/10.32744/PSE.2020.6.22

Cebrián, G., Junyent, M., & Mulà, I. (2021). Current practices and future pathways towards competencies in education for sustainable development. *Sustainability*, *13*(16), 8733. https://doi.org/10.3390/su13168733

Chawla, L. (2020). Childhood nature connection and constructive hope: A review of research on connecting with nature and coping with environmental loss. *People and Nature*, *2*(3), 619–642. https://doi.org/10.1002/pan3.10128

Cocks, S., & Simpson, S. (2015). Anthropocentric and ecocentric: An application of environmental philosophy to outdoor recreation and environmental education. *Journal of Experiential Education*, *38*(3), 216–227. https://doi.org/10.1177/1053825915571750

de Freitas Netto, S. V., Sobral, M. F. F., Ribeiro, A. R. B., & Soares, G. R. D. L. (2020). Concepts and forms of greenwashing: A systematic review. *Environmental Sciences Europe*, *32*(1), 19. https://doi.org/10.1186/s12302-020-0300-3

Florencio da Silva, R., Torres-Rivera, A. D., Alves Pereira, V., Regis Cardoso, L., & Becerra, M. J. (2023). Critical environmental education in Latin America from a socio-environmental perspective: Identity, territory, and social innovation. *Sustainability*, *15*(12), 9410. https://doi.org/10.3390/su15129410

Fraijo-Sing, B., Tapia-Fonllem, C., & Corral-Verdugo, V. (2013). Designing environmental education programs: Modeling pro-environmental competency. In *Values in Sustainable Development* (pp. 244–254). Routledge. https://doi.org/10.4324/9780203080177

Guerra, J., Schmidt, L., & Prata, L. (2022). Sustainable development goals in Portuguese speaking countries: The perspective of environmental education experts. In W. Leal Filho, D. G. Vidal, M. A. P. Dinis, & R. C. Dias (Eds.), *Sustainable Policies and Practices in Energy, Environment and Health Research: Addressing Cross-Cutting Issues* (pp. 39–59). Springer International Publishing. https://doi.org/10.1007/978-3-030-86304-3_3

Hadjileontiadou, S. J., Plastra, E., Toumpas, K., Kyprioti, K., Mandiliotis, D., Barroso, J., & Hadjileontiadis, L. J. (2013, 2014). On enhancing disabled students' accessibility in environmental education using ICT: The MusicPaint case. Universal Access in Human-Computer Interaction. Universal Access to Information and Knowledge, Cham.

Hurtado Loaiza, C. V. (2024). Hacia una Educación Ambiental Transformadora en un Contexto Rural de Colombia. *European Public & Social Innovation Review*, *9*, 1–18. https://doi.org/10.31637/epsir-2024-1084

Hyseni Spahiu, M., Korca, B., & Lindemann-Matthies, P. (2014). Environmental education in high schools in Kosovo: A teachers' perspective. *International Journal of Science Education*, *36*(16), 2750–2771. https://doi.org/10.1080/09500693.2014.933366

Igei, K., Kurokawa, H., Iseki, M., Kitsuki, A., Kurita, K., Managi, S., Nakamuro, M., & Sakano, A. (2024). Synergistic effects of nudges and boosts in environmental education: Evidence from a field experiment. *Ecological Economics*, *224*, 108279. https://doi.org/10.1016/j.ecolecon.2024.108279

Jaime, M., Salazar, C., Alpizar, F., & Carlsson, F. (2023). Can school environmental education programs make children and parents more pro-environmental? *Journal of Development Economics*, *161*, 103032. https://doi.org/10.1016/j.jdeveco.2022.103032

Junior, S. S. B., Merlo, E. M., Freire, O. B. L., Da Silva, D., & Quevedo-Silva, F. (2016). Effect of environmental concern and skepticism in the consumption green products in Brazilian retail. *Revista ESPACIOS*, *37*(2), 13.

Kiraz, A., & Danju, İ. (2025). 17—Integration of digital education into environmental education: Raising environmental awareness through the media. In F. Al-Turjman, F. Altinay, & Z. A. Gazi (Eds.), *Artificial Intelligence of Things (AIoT)* (pp. 193–202). Morgan Kaufmann. https://doi.org/10.1016/B978-0-443-26482-5.00018-3

Kopnina, H. (2020). Education for the future? Critical evaluation of education for sustainable development goals. *The Journal of Environmental Education*, *51*(4), 280–291. https://doi.org/10.1080/00958964.2019.1710444

Leal Filho, W., Lange Salvia, A., Pretorius, R. W., Londero, L. L., Manolas, E., Alves, F., Azeiteiro, U., Rogers, J., Shiel, C., & Do Paco, A. (2020). *Universities as Living Labs for Sustainable Development—Supporting the Implementation of the Sustainable Development Goals*. Springer Cham. https://doi.org/10.1007/978-3-030-15604-6

Lee, E., & Kim, N. (2025). Dynamics of school environmental education policy through systems thinking: A case study of carbon-neutral model schools in Korea. *Environmental Education Research*, 1–17. https://doi.org/10.1080/13504622.2024.2435508

Lozano, R., Barreiro-Gen, M., Lozano, F. J., & Sammalisto, K. (2019). Teaching sustainability in European higher education institutions: Assessing the connections between competences and pedagogical approaches. *Sustainability*, *11*(6), 1602. https://doi.org/10.3390/su11061602

Ma, L., Shahbaz, P., Haq, S. U., & Boz, I. (2023). Exploring the moderating role of environmental education in promoting a clean environment. *Sustainability*, *15*(10), 8127. https://doi.org/10.3390/su15108127

Mendonça, S., Piovesana, G. F., & Pissolito, V. (2025). Geoethics and sustainability: Addressing challenges in environmental education for achieving the SDGs. *Sustainability*, *17*(2), 574. https://doi.org/10.3390/su17020574

Menton, M., Larrea, C., Latorre, S., Martinez-Alier, J., Peck, M., Temper, L., & Walter, M. (2020). Environmental justice and the SDGs: From synergies to gaps and contradictions. *Sustainability Science*, *15*(6), 1621–1636. https://doi.org/10.1007/s11625-020-00789-8

Redman, F., Erin, W., & Arnim Redman, A. (2018). Continuing professional development in sustainability education for K-12 teachers: Principles, programme, applications, outlook. *Journal of Education for Sustainable Development*, *12*, 59–80. https://doi.org/10.1177/2455133318777182

Reis, É. V. B., Kokke, M., & Thomé, R. (2024). Risk management integrated with environmental education and the application of fuzzy logic. *Veredas do Direito*, *21*, e212594. https://doi.org/10.18623/rvd.v21.2594-ing

Rieckmann, M., & Barth, M. (2022). Educators' competence frameworks in education for sustainable development. In P. Vare, N. Lausselet, & M. Rieckmann (Eds.), *Competences in Education for Sustainable Development: Critical Perspectives* (pp. 19–26). Springer International Publishing. https://doi.org/10.1007/978-3-030-91055-6_3

Rousell, D., & Cutter-Mackenzie-Knowles, A. (2020). A systematic review of climate change education: Giving children and young people a 'voice' and a 'hand' in redressing climate change. *Children's Geographies*, *18*(2), 191–208. https://doi.org/10.1080/14733285.2019.1614532

Sonetti, G., Barioglio, C., & Campobenedetto, D. (2020). Education for sustainability in practice: A review of current strategies within Italian universities. *Sustainability*, *12*(13), 5246. https://doi.org/10.3390/su12135246

Ugolini, F., & Pearlmutter, D. (2022). Environmental education and location-based technologies for investigating urban threats and promoting nature-based solutions. In C. Vasconcelos & C. S. C. Calheiros (Eds.), *Enhancing Environmental Education Through Nature-Based Solutions* (pp. 413–432). Springer International Publishing. https://doi.org/10.1007/978-3-030-91843-9_24

Valenti, M. W., de Oliveira, H. T., & Logarezzi, A. J. M. (2017). Exclusory and transformative dimensions of adult environmental education in two Brazilian protected areas. *Environmental Education Research*, *23*(5), 675–686. https://doi.org/10.1080/13504622.2015.1077503

Veidemane, A. (2022). Education for sustainable development in higher education rankings: Challenges and opportunities for developing internationally comparable indicators. *Sustainability*, *14*(9), 5102. https://doi.org/10.3390/su14095102

Velempini, K., & Maruatona, T. (2025). A critical review of UNESCO's influence on lifelong learning and environmental education delivery in Botswana. *International Journal of Lifelong Education*, 1–22. https://doi.org/10.1080/02601370.2025.2497411

Wamsler, C., Osberg, G., Osika, W., Herndersson, H., & Mundaca, L. (2021). Linking internal and external transformation for sustainability and climate action: Towards a new research and policy agenda. *Global Environmental Change*, *71*, 102373. https://doi.org/10.1016/j.gloenvcha.2021.102373

Whitburn, J., Linklater, W., & Abrahamse, W. (2020). Meta-analysis of human connection to nature and proenvironmental behavior. *Conservation Biology*, *34*(1), 180–193. https://doi.org/10.1111/cobi.13381

Yadav, S. (2023). Cultivating sustainable behavior and critical consciousness towards the environment. In *The Impact of Climate Change and Sustainability Standards on the Insurance Market* (pp. 333–348). https://doi.org/10.1002/9781394167944.ch21

Section III

Methodologies and Pedagogical Strategies

7 Innovative Methodological Approaches

7.1 EXPERIENTIAL LEARNING AND OUTDOOR EDUCATION

7.1.1 Theoretical Foundations of Experiential Learning

Experiential learning constitutes a foundational pedagogical approach in environmental education, facilitating knowledge construction through direct engagement with environmental phenomena and better connect with the nature they want to understand and protect. This approach is rooted in John Dewey's philosophy of experiential education, subsequently formalized by David Kolb into a cyclical model of experiential learning.

Pappo et al. (2025) showed that experiential strategies are effective for teaching about the environment through examples. The results indicate that local examples are not always necessary to learn general concepts. However, they also showed that if activities relate to real-world situations (such as using WebQuests in real contexts), students learn and experience more. This highlights an important point: Real-world learning experiences are very useful.

According to Powell and Wells (2002) and Olsson et al. (2022), learning is a continuous cycle. First, learners have an experience. They then reflect on it. Next, they connect it. Finally, they use what they have learned to do new things. In environmental education, this cycle is observed when students engage in field activities, analyze what they observe, relate it to theory, and use what they have learned to understand or change other environmental situations.

Lhota et al. (2024) studied education for conservation as an evolving international program and a 14-year environmental education curriculum, illustrating the practical application of these principles. Their program evolved "from an environmental education program in Balikpapan Bay, Indonesian Borneo, after its critical evaluation." The researchers realized that "to achieve this goal, we would need to work with students from the age of 4 and continue educating them for many years." The program has proven to be "attractive to teachers and has spread from Kalimantan to other Indonesian islands and countries outside Asia."

7.1.2 Outdoor Education Methodologies

Outdoor education represents a specific and powerful application of experiential learning that uses natural environments as living classrooms, where students can observe, experience, and interact directly with the ecological systems they study.

DOI: 10.1201/9781003779667-10

This educational modality transcends simple knowledge transmission, facilitating transformative experiences that can profoundly influence participants' relationships with the natural world.

Basu et al. (2024) described an international experiential learning program with an interdisciplinary approach to environmental education for graduate students. Their cohort-based program worked with "three cohorts of graduate scholars from sub-Saharan Africa." The findings demonstrated that "an interdisciplinary cohort approach is beneficial for accelerating research capacity to address complex problems such as climate change." In addition, "participating scholars benefited from cross-learning, research dissemination, community engagement, network development, and improved soft skills."

The most effective outdoor education methodologies typically include components such as field orientation and navigation, systematic naturalist observation, participatory environmental monitoring, and practical conservation projects. These activities allow students to develop not only conceptual knowledge of ecological processes but also practical skills for environmental research and competencies for effective action in natural contexts.

Casillas-Zapata and Adame-Rivera (2024), in their study of playing in parks to promote environmental care in childhood, examine the relationship between "pro-environmental attitudes and behaviors with children's level of contact with nature and their visits to urban green spaces designed for play." Their mixed methodology found "a significant relationship between pro-environmental attitudes and children's visits to parks." The authors concluded that "parks can be considered outdoor play spaces and children have the potential to develop pro-environmental attitudes."

7.1.3 Designing Meaningful Immersive Experiences

The effectiveness of outdoor learning experiences depends critically on systematic instructional design. The quality of the location and the activities carried out greatly influence students' learning. For these experiences to be effective, it is important to choose the right location, organize activities logically, and prepare participants. It is also crucial to reflect on what has been done to help students understand and connect it to broader concepts.

In this regard, Lozano-Lucia and Talavera-Ortega (2024), who focused on environmental education for young children, offer valuable insights for designing these activities. Although there were slight differences, teachers in Spain and Costa Rica tended to view environmental issues broadly as interconnected. However, they also emphasized the need to improve teacher training from the outset so that this perspective can be translated into well-organized, useful, and educational outdoor activities. This study revealed the importance of valuing natural spaces as a central component of effective curriculum design.

The most impactful immersive experiences typically incorporate elements of appropriate challenge, guided discovery, and structured reflection, enabling participants to build a meaningful personal understanding through direct interactions with natural systems. These experiences may also include community service or hands-on conservation that connects individual learning to tangible contributions to environmental well-being.

Küçükaydın and Ulum (2023) on the effects of Web 2.0-supported environmental education on self-efficacy and environmental awareness found that while "Web 2.0-supported environmental education had no effect on environmental awareness," it did result in "a statistically significant difference in the perception of academic competence." This finding suggests that different experiential modalities may have varying impacts across dimensions of environmental learning.

7.1.4 Assessing Impacts on Attitudes and Behaviors

Measuring the impacts of outdoor learning experiences presents unique challenges because many of the most important outcomes, such as changes in values, emotional connection to nature, and long-term commitment to environmental behaviors, are difficult to quantify and may not manifest over extended periods.

Torralba-Burrial and Dopico (2023), promoting the sustainability of artisanal fishing through environmental education using game-based learning, illustrate systematic impact assessment. Their educational experience in promoting artisanal fishing was developed in northwestern Spain, where more than a thousand primary school students (aged 6–12) participated. Following a qualitative methodology, they analyzed the games they generated themselves, their formative evaluation, and the results of their implementation in schools and informal education events.

The most rigorous evaluation strategies employ multiple data sources and perspectives rather than relying on a single instrument. This can include directly observing behaviors in the field, conducting in-depth interviews, analyzing written reflections, and following up with participants over time to identify gradual changes in their behavior. In addition, these evaluations can incorporate transfer measures, that is, reviewing whether what was learned in outdoor experiences is later reflected in attitudes and behaviors in other contexts (e.g., at home, school, or in the community).

7.1.5 Integration with Formal Curricula

One of the best ways to increase the impact of outdoor education is to intentionally integrate it into the formal curriculum. When this integration is done correctly, greater pedagogical coherence is achieved: Activities outside the classroom are not isolated experiences but rather complement and reinforce what students learn in traditional classes.

Hunter and Richmond (2022) analyzed the theoretical diversity and inclusivity in science and environmental education research, providing conceptual frameworks for this integration. Their analysis proposes "a view of theoretical inclusivity that promises a rich and robust research landscape for both EEI [environmental education research] and SER [science education research] through the deliberate inclusion of non-Western theories." This perspective suggests that effective integration requires openness to multiple forms of knowledge and pedagogical approaches to teaching.

The most effective integration typically requires close collaboration between classroom educators and outdoor education specialists, the development of curriculum materials that connect field experiences to specific learning objectives, and assessment systems that capture the unique benefits of experiential learning.

7.2 PARTICIPATORY AND COMMUNITY-BASED METHODOLOGIES

7.2.1 Principles of Authentic Participation

Participatory methodologies in environmental education are based on the recognition that the most transformative learning processes emerge when participants take on active roles as co-researchers, co-designers, and co-implementers of their educational experiences. These approaches transcend traditional models of knowledge transmission to create horizontal learning spaces in which different forms of expertise can contribute to the collective construction of environmental understanding.

Calonge et al. (2022), in their analysis of training and dissemination about the environment as keys to promoting the abiotic component of environmental education, emphasize that "compulsory education is essential to offering an integrated and multidisciplinary vision of environmental education because it is aimed at all citizens." However, it also recognizes that "dissemination activities and strategies such as specialized mass media, science museums, and natural parks, among many others, are very useful, but they only reach a small part of society."

The principles of authentic participation include recognizing and appreciating different forms of knowledge, ensuring an equitable distribution of power in decision-making, maintaining transparency in objectives and methods, and committing to practical action that generates tangible benefits for participants and their communities. These principles require environmental educators to develop skills as facilitators, rather than as transmitters of information.

Yu and Gu (2022) analyzed the factors influencing the political environment of Chinese environmental education development and the reform trend, in combination with the history of environmental education. They provide insights into the importance of the political context for participatory methodologies. Their analysis emphasizes that "the world's major environmentally developed powers have entered the wave of environmental education reform," suggesting that participatory approaches must consider broader political and cultural frameworks.

7.2.2 Community Action Research Approaches

Community action research is a specific participatory methodology that combines learning processes, systematic research, and collective action to address real environmental problems that affect specific communities. This approach recognizes that communities possess valuable knowledge about their local environments and that the most effective research emerges when external technical expertise is combined with local knowledge.

Huo et al. (2020) focused on integrating environmental education regarding air pollution into the Personal and Social Responsibility Teaching Model in physical education illustrate innovative approaches to action research. The program incorporated air quality topics, such as PM2.5 particles, acid rain, and smog, into physical education activities so that learning occurred while students were moving and participating. The intervention was carried out in several stages (pre-class participation, mindful conversation, physical activity, group meetings, moments of introspection, and closing activities), with the aim of helping students not only identify and prevent air pollution but also develop a deeper sense of environmental responsibility. Specifically, the aim was to strengthen values such as respect, effort, and cooperation while promoting autonomy and a willingness to help others as a basis for caring for the environment.

Community action research processes typically include stages of participatory diagnosis, in which communities identify and prioritize environmental issues, collectively plan research and action, collaboratively implement activities, and reflectively evaluate to inform subsequent cycles of action and research.

7.2.3 Local Capacity Building

The fundamental objective of participatory methodologies is to develop local capacities that can sustain environmental learning and action processes independently of external intervention. Capacity building includes developing technical skills for environmental research and monitoring, organizational skills for community coordination, and political capacities for advocacy and participation in decision-making processes.

Krpálek et al. (2020) studied the environmental education for tourism, finance, and management students in the Czech Republic, examining capacity building in higher education. Their analysis of 3,680 students found that the study program influenced the inclusion of environmental components. For the tourism program, the environmental component is developed through work placements; however, qualitative research on the main subjects taught found that environmental aspects were not included in the curriculum.

The most effective strategies for local capacity building include peer mentoring programs, community experience exchanges, the creation of technical support networks, and the development of documentation and systematization systems that allow communities to capture and share their learning.

7.2.4 Integration of Traditional and Scientific Knowledge

One of the most valuable contributions of participatory methodologies is their potential to facilitate genuine dialogue between different knowledge systems, particularly between traditional community knowledge and formal scientific knowledge systems. This integration can generate richer and more contextually grounded cognitive syntheses than either of these knowledge systems can produce independently.

Fernández-Llamazares et al. (2021), in their scientists' warning to humanity on threats to indigenous and local knowledge systems, demonstrate the critical importance of integrating traditional ecological knowledge with formal scientific approaches in environmental education. Their analysis found that indigenous and local knowledge systems contain invaluable information about ecological processes, species interactions, and sustainable resource management practices that have been refined over generations. However, these knowledge systems face unprecedented threats from cultural erosion, environmental degradation, and educational systems that privilege Western scientific paradigms. The authors concluded that environmental education programs must actively create spaces for respectful dialogue between knowledge traditions, recognizing that the integration of traditional and scientific knowledge produces more comprehensive and culturally appropriate understandings of complex socioecological systems.

The most effective processes for integrating knowledge include methodologies such as participatory knowledge mapping, where different stakeholder groups contribute their perspectives on socioecological systems; structured knowledge dialogues that facilitate respectful exchanges between bearers of different knowledge traditions; and collaborative research projects where formal scientific methodologies are combined with traditional observation and experimentation systems.

7.2.5 Participatory Evaluation Methodologies

Evaluating programs using participatory methodologies requires approaches that capture not only individual learning outcomes but also changes in collective capacities, the strengthening of social capital, and transformations in community power dynamics. These evaluation approaches must be consistent with the participatory principles that guide the implementation of programs.

Birdsall and France (2018) developed an analytical tool to untangle environmental education resources and provide methodological frameworks for participatory evaluation. Their tool is based on "a socially acute questions (SAQ) approach and educational configurations that teachers use when implementing this approach." Resource analysis revealed that educational materials can have very different "epistemic stances" and "didactic approaches," underscoring the importance of evaluation that captures this diversity.

The most promising participatory evaluation methodologies include community self-evaluation, in which communities develop criteria and processes to evaluate their programs; peer evaluation among communities implementing similar initiatives; and more meaningful development evaluation approaches that prioritize changes that participants consider most important.

7.3 GAMIFICATION AND STORYTELLING IN ENVIRONMENTAL EDUCATION

7.3.1 Theoretical Foundations of Educational Gamification

Gamification in environmental education is based on the recognition that games have intrinsic characteristics that can facilitate deep learning, sustained engagement, and intrinsic motivation, often in ways that are difficult to achieve through traditional educational methodologies. These elements include clear goal setting, immediate feedback, structured progression, meaningful choices, and opportunities for experimentation and iteration in the learning process.

Ouariachi et al. (2020) conducted a systematic review searching for best practices in gamification approaches for education and engagement on pro-environmental behaviors. Their analysis identified key design principles that make environmental games effective, including the use of narrative framing, social interaction mechanics, and feedback systems that link player decisions to environmental consequences. The authors found that gamification approaches are most successful when they combine intrinsic motivation elements with meaningful environmental content, and when they provide opportunities for players to transfer gaming experiences to real-world pro-environmental actions.

The most effective game elements in environmental education contexts include exploration mechanics that allow students to discover environmental information autonomously, scoring systems that reward environmentally responsible behaviors, collaborative challenges that require teamwork to solve complex environmental problems, and immersive narratives that connect individual actions to broader environmental consequences.

Debele and Plevyak (2017), on Ethiopian environmental education from a utopian perspective, compare "the perspectives of two third-grade students in a private Ethiopian school with an anti-utopian and utopian perspective on the environment." The study found that "the difference in the children's social and environmental perspectives showed that their natural disposition to interact with nature holistically should be considered in the development of an effective environmental education curriculum."

7.3.2 Environmental Narratives and Meaning Construction

Storytelling in environmental education leverages the fundamental human capacity to organize experiences and knowledge through narrative structures, making complex information more accessible, memorable, and relevant. Effective environmental narratives can connect abstract global issues with concrete local experiences, facilitate emotional identification with people, species, and ecosystems affected by environmental problems, and provide a frame of reference for individual and collective action.

Veland et al. (2018) examined the transformative role of storytelling in realizing 1.5°C climate futures. Their analysis demonstrated that narratives play a crucial role in making abstract environmental challenges—such as climate change—personally meaningful and actionable. The authors argued that effective environmental storytelling creates spaces for "reimagining alternative futures," enabling audiences to envision and emotionally connect with sustainable pathways. Their findings suggest that narrative approaches in environmental education are most powerful when they combine scientific accuracy with emotional resonance, cultural relevance, and a sense of agency that empowers listeners to see themselves as active participants in environmental solution.

However, when considering how stories help with environmental education, it is best to tell stories that are relatable and true. The ones that work best have characters people can identify with, environmental problems that mirror everyday ones, endings that show something can be done, and messages that connect the story to broader environmental ideas. These stories can be told in many ways, from books and plays to videos, interactive multimedia, and virtual reality (VR) experiences.

7.3.3 Designing Environmental Gaming Experiences

Designing game experiences for environmental education involves balancing what makes a game fun and engaging with what enables it to truly teach and bring about lasting changes in how people think and act. Achieving this balance presents significant design challenges: Excessive emphasis on didactic content can undermine the intrinsic motivation that makes gamification effective as a learning tool and internal motivation that makes it such a powerful learning tool.

Stern et al. (2014) evaluated environmental education programs in the New Millennium and examined what is measured and what has been learned through a systematic review of the literature. Their analysis of peer-reviewed studies published between 1999 and 2010 found that "while the review generally supports consensus-based best practices, we also identified additional themes that can drive positive outcomes, including the provision of holistic experiences and the characteristics and delivery styles of environmental educators."

The most important principles for environmental game design include establishing clear, measurable learning objectives, incorporating authentic and up-to-date environmental content, designing game mechanics that reflect real ecological processes and relationships, developing feedback systems that link player actions to environmental consequences, and including opportunities for reflection and transferring learning to real-world contexts.

7.3.4 Emerging Technologies for Gamification

Emerging technologies have significantly expanded the possibilities of creating immersive, interactive, and personalized environmental gaming experiences. These technologies include VR and augmented reality (AR), artificial intelligence (AI), big data analytics, the Internet of Things, and social gaming platforms that facilitate collaboration among geographically distributed players.

Petersen et al. (2020) investigated how to optimize immersive virtual learning in climate change education through virtual field trips. Their experimental study compared different design approaches for virtual environmental experiences and found that immersive VR field trips significantly enhanced students' learning outcomes about climate change compared to traditional instructional methods. The researchers identified key design principles for effective virtual environmental experiences, including the importance of embodied interaction, guided exploration, and reflective pauses that allow learners to process and integrate new information. These findings demonstrate how emerging technologies can create powerful gamified learning environments that transport students to otherwise inaccessible ecosystems and environmental scenarios.

The most promising applications of emerging technologies include ecosystem simulations that allow players to experiment with different environmental management strategies, AR games that overlay environmental information on real environments, branching citizen-science platforms that reward contributions to real environmental research, and AI systems that can personalize gaming experiences based on learning styles and individual progress.

7.3.5 Impact Assessment of Playful Approaches

Assessing the educational impact of gamification and storytelling approaches presents unique challenges because these approaches can foster learning outcomes such as intrinsic motivation, emotional engagement, and creative transfer, which are difficult to measure with traditional assessments. In addition, the most important effects may manifest over extended periods after the initial gaming experience.

Viteri et al. (2014) on children's recall and motivation for an environmental education video with supporting teaching materials examined differences in recall among high-, average-, and low-performing fifth-grade students for an environmental education video with supporting teaching materials. Their study also assessed "the motivational level of all students." The findings suggest "comparative recall results" for all students and high motivation for the designed video and teaching materials.

The most effective assessment methodologies for playful approaches combine multiple sources of evidence, including observations of behavior during play, analyses of play data that reveal decision-making patterns, interviews that explore participants' subjective experiences, and transfer assessments that examine whether play learning influences attitudes and behaviors in other contexts.

7.4 INTEGRATION OF DIGITAL TECHNOLOGIES AND EMERGING TOOLS

7.4.1 Digital Revolution in Environmental Education

The integration of digital technologies into environmental education has fundamentally transformed the possibilities for creating learning experiences that are more accessible, interactive, personalized, and effective than traditional methods of teaching. This transformation has been accelerated by developments in mobile computing, ubiquitous connectivity, environmental sensors, and online collaboration platforms that can connect learners with real-time environmental information and global communities of practice.

Ardoin et al. (2020) conducted a systematic review of environmental education outcomes for conservation and found that digital and non-digital educational interventions alike can produce significant positive outcomes across cognitive, affective, and behavioral domains. Their analysis of over 100 peer-reviewed studies revealed that the most effective programs integrate multiple pedagogical approaches, leverage technology to extend learning beyond traditional settings, and connect educational experiences to authentic conservation challenges. These findings provide a strong evidence base for the digital revolution in environmental education, demonstrating that technology-enhanced approaches can maintain the depth and transformative potential of traditional methods while dramatically expanding accessibility and reach.

The most impactful digital technologies in environmental education include adaptive learning platforms that can personalize content and educational sequences based on individual needs and progress, simulation tools that allow experimentation with complex environmental systems, AR applications that overlay environmental information on real environments, and collaborative monitoring systems that involve citizens in the collection and analysis of environmental data.

7.4.2 Interactive Learning Platforms

The development of interactive learning platforms has democratized access to high-quality environmental education experiences, allowing students in remote locations or with limited resources to use sophisticated simulations, up-to-date environmental databases, and collaborate with experts and peers worldwide.

Rousell and Cutter-Mackenzie-Knowles (2020) conducted a systematic review of climate change education research and identified the critical role of interactive platforms in giving children and young people a "voice" and a "hand" in redressing climate change. Their analysis found that the most effective digital learning platforms are those that move beyond passive information delivery to create spaces where learners can actively engage with environmental data, collaborate with peers, and develop their own responses to environmental challenges. The review emphasized that interactive platforms should incorporate student agency, creative expression, and opportunities for authentic participation in environmental decision-making.

The most effective platforms typically include features such as rich multimedia content that accommodates different learning styles, collaboration tools that facilitate teamwork and the exchange of perspectives, formative assessment systems that provide immediate feedback and personalized guidance, and connections to real-world resources that extend learning beyond digital environments.

7.4.3 VR and AR in Environmental Education

VR and AR technologies have opened up unprecedented possibilities for creating immersive environmental education experiences that can transport students to distant ecosystems, allow them to manipulate complex environmental variables, and visualize processes that occur at temporal or spatial scales that are normally beyond direct human experience.

Markowitz et al. (2018) and Han (2021) demonstrated that immersive VR field trips significantly facilitate learning about climate change. Their experimental study found that participants who experienced ocean acidification through an immersive VR simulation showed greater knowledge gains, stronger concern for the environment, and more positive attitudes toward environmental action compared to those who received the same information through traditional video. Building on these findings, Han (2021) investigated the impact of immersive virtual field trips on elementary students' perceptions and found that VR experiences generated high levels of engagement, enhanced spatial understanding of environmental phenomena, and created memorable learning experiences that students could recall and apply in subsequent lessons. Together, these studies provide compelling evidence that VR and

AR technologies can create powerful educational experiences that complement and extend traditional outdoor education.

The most promising applications of VR and AR in environmental education include virtual tours of ecosystems that may be inaccessible due to distance, cost, or conservation restrictions; interactive simulations that allow students to experience the consequences of different environmental decisions; visualizations of complex environmental data that can make abstract patterns more understandable; and multiple perspective experiences that allow students to experience environmental systems from the perspectives of different species or stakeholders.

7.4.4 Citizen Science and Collaborative Monitoring

The integration of digital technologies has enabled the development of citizen-science programs that engage students and communities in authentic environmental research while providing hands-on learning experiences that connect education to practical action for environmental understanding and protection.

Fraisl et al. (2022) published a comprehensive review of citizen science in environmental and ecological sciences in Nature Reviews Methods Primers. Their analysis demonstrated that citizen-science projects have expanded dramatically in scope and sophistication, enabled by digital technologies that allow volunteers to collect, share, and analyze environmental data at unprecedented scales. The review identified key design principles for effective citizen-science platforms, including user-friendly data collection interfaces, real-time feedback mechanisms that show participants how their contributions fit into larger research efforts, automated data quality assurance systems, and community features that foster social learning and sustained engagement. Their findings confirm that citizen science simultaneously advances scientific research and provides powerful learning experiences that connect participants to authentic environmental inquiry.

The most effective citizen-science projects provide participants with user-friendly digital tools for data collection, platforms for sharing observations with broader communities of researchers and citizens, access to data analysis and visualizations that make individual contributions more meaningful, and opportunities to interpret the results and develop recommendations for action.

7.4.5 AI and Learning Personalization

Developments in AI have created opportunities to personalize environmental education experiences to accommodate individual differences in learning styles, prior knowledge, interests, and motivation. These technologies can analyze learning behavior patterns in real time and adapt content, sequences, and methodologies to optimize educational outcomes for students.

Holmes et al. (2023), in their UNESCO publication on AI in education, provided a comprehensive framework for understanding how AI technologies can transform educational practices across disciplines, including environmental education. Their analysis identified several key applications of AI that are particularly relevant for environmental learning: Intelligent tutoring systems that adapt content difficulty and pace to individual learner needs, natural language processing tools that can facilitate dialogue and reflection on complex environmental topics, machine learning algorithms that can identify patterns in student engagement and predict learning difficulties, and recommendation engines that can suggest personalized learning pathways based on individual interests and prior knowledge. The authors emphasized that while AI offers unprecedented opportunities for personalizing education, its implementation must be guided by ethical principles that ensure equity, transparency, and human oversight.

The most promising applications of AI in environmental education include intelligent tutorial systems that can provide personalized guidance and adaptive feedback, sentiment analysis tools that can identify emotional engagement and adjust experiences accordingly, recommendation systems that can suggest resources and activities based on individual interests and progress, and conversational agents that can facilitate dialogue and reflection on complex environmental issues.

7.4.6 Ethical and Accessibility Challenges

The integration of digital technologies into environmental education also presents significant challenges related to equity of access, data privacy, technological dependence, and alignment with the environmental values of sustainability and social justice, which are central to effective environmental education.

Menton et al. (2020) examined the relationship between environmental justice and the Sustainable Development Goals (SDGs), identifying both synergies and contradictions. Their analysis is directly relevant to the ethical challenges of digital technology integration in environmental education, as they demonstrated that environmental benefits and burdens are unevenly distributed across communities defined by race, class, gender, and geography. Applied to digital environmental education, their findings underscore that the digital divide can reproduce and amplify existing environmental injustices: Communities most affected by environmental degradation are often those with the least access to digital educational technologies. This highlights the ethical imperative for environmental education programs to address digital equity as a fundamental component of environmental justice.

The most responsible approaches to technology integration require careful attention to issues such as the digital divide, which can exclude communities with limited access to technology, privacy policies that protect students' personal information, universal design that accommodates different abilities and needs, and the environmental sustainability of the technologies used in educational programs (Wamsler et al., 2021).

7.5 CONCLUSION

7.5.1 Synthesis of Methodological Innovations

The analysis of innovative methodological approaches in environmental education presented in this chapter reveals a constantly evolving field that has developed a rich diversity of pedagogical strategies that can address the complexity of contemporary environmental challenges. Each methodological approach examined experiential learning and outdoor education, participatory and community-based methodologies, gamification and storytelling, and the integration of digital technologies, offering unique strengths that can support specific environmental education goals.

Ardoin et al. (2020), in their systematic review of over 100 peer-reviewed studies, confirmed that programs integrating multiple methodological approaches—combining experiential, participatory, digital, and narrative elements—consistently produced stronger outcomes across cognitive, affective, and behavioral domains than those relying on single methodological strategies. This finding underscores the importance of methodological diversity and innovation in environmental education.

Experiential learning and outdoor education have proven particularly effective in developing deep emotional connections with natural systems and facilitating holistic understandings of complex ecological processes. Participatory methodologies have unique strengths in integrating diverse forms of knowledge, building local capacity, and connecting learning to community action. Gamification and storytelling approaches have shown strong potential to motivate engagement, personalize experiences, and make complex environmental concepts more accessible.

7.5.2 Critical Factors for Effective Implementation

The analysis identified several critical factors that determine the effectiveness of innovative methodological approaches in various contexts. First, alignment between specific methodologies and particular learning objectives is fundamental. Different approaches are more effective for different types of educational outcomes, and the selection of methodologies should be guided by a clear understanding of what is to be achieved in the study (Redman et al., 2018).

Fraisl et al. (2022) demonstrated through their comprehensive review that citizen-science platforms represent a successful model of methodological synergy, integrating digital technologies with participatory methodologies, experiential learning through authentic data collection, and gamification elements through progress tracking and community recognition. Their analysis confirms first that the most effective environmental education programs emerge at the intersection of multiple methodological traditions.

Second, educator preparation and capacity building are crucial for successful implementation. Innovative methodological approaches often require pedagogical skills that go beyond traditional content delivery, including facilitating group processes, designing experiences, conducting formative assessments, and integrating technology. Third, contextual adaptation appears to be essential for its effectiveness. Methodological approaches that are effective in certain contexts may require significant modifications to function in different cultural, socioeconomic, and geographic environments.

7.5.3 Synergies between Methodological Approaches

One of the most important conclusions of this analysis is that the most effective methodological approaches often combine elements of multiple strategies rather than relying on single, isolated approaches. The integration of different methodologies can create synergies that amplify the strengths of each approach while mitigating their limitations.

Österlind (2005) researched concept formation in environmental education by studying the work of 14-year-old students on the greenhouse effect and on ozone depletion. Their case study "describing the work of three students at the upper level of compulsory school" found that "the need for certain domain-specific knowledge becomes apparent; for example, understanding concepts such as photosynthesis, radiation, and catalyst.

For example, programs that combine immersive field experiences with digital tools for documentation and analysis can provide both an emotional connection to natural systems and technical skills for rigorous inquiry. Initiatives that integrate participatory methodologies with gamification elements can foster community empowerment and sustained motivation (Brundiers et al., 2021).

7.5.4 Directions for Future Innovation

The analysis suggests several promising directions for future methodological innovations in environmental education. First, the development of hybrid approaches that combine physical and digital experiences in more seamless and integrated ways represents an important frontier for exploration. Emerging technologies such as mixed reality, the Internet of Things, and AI can facilitate experiences that transcend the traditional limitations of online and offline learning (Kopnina, 2020).

Chatzifotiou (2005) examined the interplay between national policy and local awareness in the implementation of environmental education in primary schools in northern Greece. The study revealed that various philosophical perspectives on nature significantly influence the awareness of environmental education, and both local and state factors play a mediating role in its implementation. This suggests that future methodological innovations should consider both the global and local dimensions of environmental issues.

Second, personalizing educational experiences based on individual differences in learning styles, cultural backgrounds, and local contexts is another important area of development. AI technologies can facilitate this personalization; however, they also require careful attention to issues of equity and inclusion.

Third, scaling innovative methodological approaches from pilot projects to systemic implementation presents significant challenges and opportunities for researchers and practitioners. This requires the development of dissemination models, technical support systems, and evaluation frameworks that sustain innovation during expansion (Chawla, 2020).

7.5.5 Imperative for Collaborative Action

The evidence presented in this chapter calls for collaborative action among researchers, educators, technologists, policymakers, and communities to accelerate the development, evaluation, and implementation of innovative methodological approaches to maximize the transformative impact of environmental education.

This collaboration must be guided by the principles of inclusion, equity, and sustainability to ensure that methodological innovations contribute to the broader goals of environmental justice and sustainable development. It also requires a commitment to rigorous evaluation and continuous improvement that can generate evidence of what works, for whom, and under what conditions.

Menton et al. (2020), in their analysis of environmental justice and the SDGs, demonstrated that collaborative action in environmental education must explicitly address issues of equity, inclusion, and justice to ensure that methodological innovations serve all communities rather than reproducing existing disparities. Their findings underscore that the collaborative imperative extends beyond methodological development to encompass fundamental questions about who benefits from and who participates in environmental education innovation.

The urgency of contemporary environmental challenges, such as climate change, biodiversity loss, and ecosystem degradation, requires collaborative action on an unprecedented scale and speed. However, evidence also shows that effective methodological innovation cannot be rushed; it requires careful experimentation, rigorous evaluation, and iterative refinement, which can take years to fully develop.

Ultimately, the future of innovative methodological approaches in environmental education depends on the field's ability to balance scientific rigor and pedagogical creativity, operational efficiency and transformational depth, and technical scalability and local relevance. Achieving this balance requires an ongoing commitment to the values of excellence, equity, and sustainability that define the best environmental education.

REFERENCES

Ardoin, N. M., Bowers, A. W., & Gaillard, E. (2020). Environmental education outcomes for conservation: A systematic review. *Biological Conservation*, *241*, 108224. https://doi.org/10.1016/j.biocon.2019.108224

Basu, O. D., Isabelle, D., & Pineau, S. (2024). Climate change and societal transformation: A cohort-based international experiential learning program with an interdisciplinary focus on environmental education for graduate students. *Environmental Education Research*, 1–23. https://doi.org/10.1080/13504622.2024.2390627

Birdsall, S., & France, B. (2018). Lifting the cover: Developing an analytical tool to unravel environmental education resources. *Environmental Education Research*, *24*(11), 1546–1564. https://doi.org/10.1080/13504622.2017.1393046

Brundiers, K., Barth, M., Cebrián, G., Cohen, M., Diaz, L., Doucette-Remington, S., Dripps, W., Habron, G., Harré, N., Jarchow, M., Losch, K., Michel, J., Mochizuki, Y., Rieckmann, M., Parnell, R., Walker, P., & Zint, M. (2021). Key competencies in sustainability in higher education—toward an agreed-upon reference framework. *Sustainability Science*, *16*(1), 13–29. https://doi.org/10.1007/s11625-020-00838-2

Calonge, A., Molina-Navarro, E., & Alfaro, P. (2022). Training and dissemination about the environment: Keys to impulse the abiotic component of environmental education. In C. Vasconcelos & C. S. C. Calheiros (Eds.), *Enhancing Environmental Education Through Nature-Based Solutions* (pp. 9–23). Springer International Publishing. https://doi.org/10.1007/978-3-030-91843-9_2

Casillas-Zapata, A. M., & Adame-Rivera, L. M. (2024). Jugar en el parque para promover el cuidado ambiental en la infancia. *Revista Latinoamericana de Ciencias Sociales, Niñez y Juventud*, *22*(3), 1–3. https://doi.org/10.11600/rlcsnj.22.3.6232

Chatzifotiou, A. (2005). National policy, local awareness: Implementing environmental education in the primary schools of northern Greece. *Environmental Education Research*, *11*(5), 503–523. https://doi.org/10.1080/13504620500169650

Chawla, L. (2020). Childhood nature connection and constructive hope: A review of research on connecting with nature and coping with environmental loss. *People and Nature*, *2*(3), 619–642. https://doi.org/10.1002/pan3.10128

Debele, M., & Plevyak, L. (2017). Ethiopian environmental education from a utopian perspective: A comparative study of two children's outlooks. *Applied Environmental Education & Communication*, *16*(4), 213–222. https://doi.org/10.1080/1533015X.2017.1322015

Fernández-Llamazares, Á., Lepofsky, D., Lertzman, K., Armstrong, C. G., Brondizio, E. S., Gavin, M. C., Lyver, P. O. B., Nicholas, G. P., Pascua, P. A., Reo, N. J., Reyes-García, V., Turner, N. J., Yletyinen, J., Anderson, E. N., Balée, W., Cariño, J., David-Chavez, D. M., Dunn, C. P., Garnett, S. C., Greening, S., S. J., Kuhnlein, H., Molnár, Z., Odonne, G., Retter, G.-B., Ripple, W. J., Sáfián, L., Bahraman, A. S., Torrents-Ticó, M., & Vaughan, M. B. (2021). Scientists' warning to humanity on threats to indigenous and local knowledge systems. *Journal of Ethnobiology*, *41*(2), 144–169. https://doi.org/10.2993/0278-0771-41.2.144

Fraisl, D., Hager, G., Bedessem, B., Gold, M., Hsing, P.-Y., Danielsen, F., Hitchcock, C. B., Hulbert, J. M., Piera, J., Spiers, H., Thiel, M., & Haklay, M. (2022). Citizen science in environmental and ecological sciences. *Nature Reviews Methods Primers*, *2*(1), 64. https://doi.org/10.1038/s43586-022-00144-4

Han, I. (2021). Immersive virtual field trips and elementary students' perceptions. *British Journal of Educational Technology*, *52*(1), 179–195. https://doi.org/https://doi.org/10.1111/bjet.12946

Holmes, W., Bialik, M., & Fadel, C. (2023). *Artificial Intelligence in Education*. UNESCO. https://doi.org/10.54675/PCSP7350

Hunter, R. H., & Richmond, G. (2022). Theoretical diversity and inclusivity in science and environmental education research: A way forward. *Journal of Research in Science Teaching*, *59*(6), 1065–1085. https://doi.org/10.1002/tea.21752

Huo, Y.-J., Shih, K.-T., & Lin, C.-J. (2020). The study on integrating air pollution environmental education into the teaching personal and social responsibility model in physical education. *IOP Conference Series: Earth and Environmental Science*, *576*(1), 012006. https://doi.org/10.1088/1755-1315/576/1/012006

Kopnina, H. (2020). Education for the future? Critical evaluation of education for sustainable development goals. *The Journal of Environmental Education*, *51*(4), 280–291. https://doi.org/10.1080/00958964.2019.1710444

Krpálek, P., Berková, K., Krpálková Krelová, K., Kubišová, A., Frendlovská, D., & Szabo, S. (2020). Environmental education in the preparation of students of tourism and finance and management in the Czech Republic. *Sustainability*, *12*(17), 6736. https://doi.org/10.3390/su12176736

Küçükaydın, M. A., & Ulum, H. (2023). The effects of Web 2.0 supported environmental education on self-efficacy belief regarding environmental education and environmental awareness. *Education and Information Technologies*, *28*(11), 14529–14551. https://doi.org/10.1007/s10639-023-11783-9

Lhota, S., Bukit, Windi L., Razafindravony, Lovasoa E., & Randrianarivelo, H. (2024). Education for conservation, and evolving international program and 14-year curriculum of environmental education. *Folia Primatologica*, *95*(4–6), 451–461. https://doi.org/10.1163/14219980-bja10031

Lozano-Lucia, O. R., & Talavera-Ortega, M. (2024). Enfoque de la educación ambiental en educación infantil: Estudio comparado en docentes de España y Costa Rica. *Revista Electrónica Educare*, *28*(2), 22–40. https://doi.org/10.15359/ree.28-2.18430

Markowitz, D. M., Laha, R., Perone, B. P., Pea, R. D., & Bailenson, J. N. (2018). Immersive virtual reality field trips facilitate learning about climate change [Original research]. *Frontiers in Psychology*, *9*. https://doi.org/10.3389/fpsyg.2018.02364

Menton, M., Larrea, C., Latorre, S., Martinez-Alier, J., Peck, M., Temper, L., & Walter, M. (2020). Environmental justice and the SDGs: From synergies to gaps and contradictions. *Sustainability Science*, *15*(6), 1621–1636. https://doi.org/10.1007/s11625-020-00789-8

Olsson, D., Gericke, N., & Boeve-de Pauw, J. (2022). The effectiveness of education for sustainable development revisited: A longitudinal study on secondary students' action competence for sustainability. *Environmental Education Research*, *28*(3), 405–429. https://doi.org/10.1080/13504622.2022.2033170

Österlind, K. (2005). Concept formation in environmental education: 14-year olds' work on the intensified greenhouse effect and the depletion of the ozone layer. *International Journal of Science Education*, *27*(8), 891–908. https://doi.org/10.1080/09500690500038264

Ouariachi, T., Li, C.-Y., & Elving, W. J. L. (2020). Gamification approaches for education and engagement on pro-environmental behaviors: Searching for best practices. *Sustainability*, *12*(11), 4565. https://www.mdpi.com/2071-1050/12/11/4565

Pappo, E., Borden, J. B., Good, R. Z., & Coenen, D. (2025). Does local context matter for case-based environmental education? *Case Studies in the Environment*, *9*(1), 2483291. https://doi.org/10.1525/cse.2025.2483291

Petersen, G. B., Klingenberg, S., Mayer, R. E., & Makransky, G. (2020). The virtual field trip: Investigating how to optimize immersive virtual learning in climate change education. *British Journal of Educational Technology*, *51*(6), 2099–2115. https://doi.org/10.1111/bjet.12991

Powell, K., & Wells, M. (2002). The effectiveness of three experiential teaching approaches on student science learning in fifth-grade public school classrooms. *The Journal of Environmental Education*, *33*(2), 33–38. https://doi.org/10.1080/00958960209600806

Redman, E., Wiek, A., & Redman, A. (2018). Continuing professional development in sustainability education for K-12 teachers: Principles, programme, applications, outlook. *Journal of Education for Sustainable Development*, *12*(1), 59–80. https://doi.org/10.1177/2455133318777182

Rousell, D., & Cutter-Mackenzie-Knowles, A. (2020). A systematic review of climate change education: Giving children and young people a 'voice' and a 'hand' in redressing climate change. *Children's Geographies*, *18*(2), 191–208. https://doi.org/10.1080/14733285.2019.1614532

Stern, M. J., Powell, R. B., & Hill, D. (2014). Environmental education program evaluation in the new millennium: What do we measure and what have we learned? *Environmental Education Research*, *20*(5), 581–611. https://doi.org/10.1080/13504622.2013.838749

Torralba-Burrial, A., & Dopico, E. (2023). Promoting the sustainability of artisanal fishing through environmental education with game-based learning. *Sustainability*, *15*(17), 12905. https://doi.org/10.3390/su151712905

Veland, S., Scoville-Simonds, M., Gram-Hanssen, I., Schorre, A. K., El Khoury, A., Nordbø, M. J., Lynch, A. H., Hochachka, G., & Bjørkan, M. (2018). Narrative matters for sustainability: The transformative role of storytelling in realizing 1.5°C futures. *Current Opinion in Environmental Sustainability*, *31*, 41–47. https://doi.org/https://doi.org/10.1016/j.cosust.2017.12.005

Viteri, F., Clarebout, G., & Crauwels, M. (2014). Children's recall and motivation for an environmental education video with supporting pedagogical materials. *Environmental Education Research*, *20*(2), 228–247. https://doi.org/10.1080/13504622.2013.771734

Wamsler, C., Osberg, G., Osika, W., Herndersson, H., & Mundaca, L. (2021). Linking internal and external transformation for sustainability and climate action: Towards a new research and policy agenda. *Global Environmental Change*, *71*, 102373. https://doi.org/10.1016/j.gloenvcha.2021.102373

Yu, T., & Gu, M. (2022). Analysis on the policy environment influencing factors of Chinese environmental education development and the reform trend in combination with environmental education history. *Journal of Environmental and Public Health*, *2022*(1), 3454709. https://doi.org/10.1155/2022/3454709

8 Curriculum Design and Contextualized Content

8.1 CURRICULUM DESIGN PRINCIPLES FOR ENVIRONMENTAL EDUCATION

8.1.1 Comprehensiveness and Systems Thinking as Curricular Foundations

Effective curriculum design in environmental education (EE) must be based on the principle of comprehensiveness, recognizing the systemic and interconnected nature of contemporary environmental issues. This comprehensiveness requires curricula to move beyond fragmented, discipline-based approaches and embrace transdisciplinary perspectives that address the complexity of socioecological systems (Kopnina, 2020).

Ali et al. (2025) present empirical evidence supporting the efficacy of integrated EE in fostering pro-environmental attitudes. Their study in a secondary school in Indonesia found that "integrated environmental education significantly improves students' environmental attitudes ($t = 10.832$, $p < 0.001$), with 12th grade students showing higher levels of environmental awareness than those in lower grades ($F = 4.73$, $p = 0.009$)."

Significantly, the study revealed that "socioeconomic status, prior environmental knowledge, and parental support significantly influenced students' pro-environmental attitudes." However, "a gap between awareness and pro-environmental behaviors persisted, emphasizing the need for experiential learning and stronger social support." These findings underscore the need for a comprehensive curriculum design that addresses not only cognitive dimensions but also social and experiential factors influencing the development of authentic environmental engagement.

Carl et al. (2025) explore the intersection of physical education and EE, offering additional insights into the curriculum's comprehensiveness. Their integrative narrative review through a physical literacy lens found that "physical literacy can harmonize with educational work by respecting the disciplinary interests of both physical education and environmental education." However, "few intervention studies translate holistic claims of physical literacy into interventions," suggesting that curricular integration requires more sophisticated methodological development for effective implementation.

8.1.2 Territorial Contextualization and Local Relevance

The principle of territorial contextualization holds that the most effective EE emerges when abstract knowledge is connected to concrete experiences in specific territories. This contextualization requires curricula to incorporate the ecological, social,

DOI: 10.1201/9781003779667-11

economic, and cultural characteristics of the local contexts in which educational programs are implemented.

The principle of territorial contextualization is further supported by research on intergenerational learning in EE. Ardoin et al. (2020) conducted a systematic review of EE outcomes and found that programs incorporating intergenerational and place-based elements consistently produced stronger and more durable learning outcomes. Their analysis revealed that effective territorial contextualization must consider not only geographical and ecological characteristics but also the social and relational dynamics that emerge when learners of different ages and backgrounds interact within specific environmental contexts. These findings suggest that curriculum designers should create opportunities for intergenerational exchange as a core component of contextualized EE.

8.1.3 Participation and Collective Knowledge Construction

Participatory curriculum design recognizes that the most transformative educational processes emerge when participants actively contribute to the construction of educational content and methodologies (Wamsler et al., 2021). This participation requires flexible curricular frameworks that can incorporate diverse knowledge, experience, and perspectives while maintaining consistency in fundamental learning objectives.

Shahat et al. (2025), whose analysis of the use of EE to develop sustainability competencies related to experimentation with environmental materials, demonstrated the effectiveness of participatory approaches. Their quasi-experimental study with 80 high school students found "a statistically significant difference between the average scores of the experimental and control groups in the post-measurement on the SCEEM scale, in favor of the experimental group."

Crucially, "there were no statistically significant differences between the average scores of the experimental group in the post-measurement and follow-up on the SCEEM scale, as the total test value reached 1.610, which is a non-significant value, indicating the continuity of the program's effect and its effectiveness." This suggests that participatory approaches can generate lasting impacts that persist beyond immediate educational interventions.

8.1.4 Progressive Sequencing and Building Complexity

Effective curriculum design must sequence content and learning experiences to support the progressive construction of environmental understanding and competency. This sequencing requires conceptual frameworks that can identify the cognitive, affective, and experiential prerequisites for different levels of environmental learning.

Mbah and Ezegwu (2024) provide important insights into curriculum sequencing in their study on decolonizing climate change and EE in Africa. Their systematic review of articles published between 2015 and 2022 found "persistent exclusion and epistemic alienation of indigenous and local knowledge (ILK) in CCEE (climate change and environmental education)."

The researchers emphasized that "the lack of recognition of ILK contributes to negatively affecting the quality of what is learned in educational settings across

Africa, as students' lived experiences in their sociocultural environments are not contextualized to improve the relatability of their learning topics." This underscores the need for effective curriculum sequencing to integrate multiple forms of knowledge from the early stages of curriculum design.

8.1.5 Formative Assessment and Continuous Improvement

The principle of formative assessment recognizes that curriculum design must include systematic mechanisms to monitor educational effectiveness and make adjustments based on evidence of student learning and feedback from the implementers. This assessment requires indicator systems that can capture multiple dimensions of environmental learning.

Polley et al. (2024) offer conceptual frameworks for evaluating threshold concepts in outdoor and EE within Australian universities. Their analysis of assessment practices in higher education programs found that "threshold concepts can offer a curriculum development framework that highlights the graduate capabilities of a higher education outdoor educator," but "do not specify how these concepts are assessed."

The researchers concluded that "authentic assessment" may be crucial for "assessing the seven threshold concepts of higher education outdoor education." They provided "five recommendations for how academics might assess" these concepts to "support the communication of graduate capabilities to employers, students, graduates, and other outdoor education stakeholders."

8.1.6 Flexibility and Methodological Innovation

Contemporary curriculum design must incorporate principles of flexibility that enable adaptation to diverse contexts, integrate emerging technologies, and experiment with innovative pedagogical methodologies (Rousell & Cutter-Mackenzie-Knowles, 2020). This flexibility requires curriculum frameworks that can balance structure and openness, as well as coherence and adaptability.

Florencio da Silva et al. (2023) emphasize the significance of adaptability in critical EE within Latin America, particularly from a socio-environmental perspective. Their research investigates the manner in which environmental issues related to justice within territories and communities confront the complexities arising from the effects of climate change. They found that "critical environmental education in Latin America reveals historical distortions, forms of colonization, and modes of production associated with the exploitation of nature that deepen extreme poverty." This evidence suggests that curricular flexibility should include the capacity to address dimensions of social justice and systemic transformation, which may require innovative pedagogical approaches.

8.2 CULTURAL AND CONTEXTUAL ADAPTATION OF CONTENT

8.2.1 Recognition of Epistemological Diversity

Effective cultural adaptation of EE curriculum content requires a deep recognition that different cultures have developed distinctive ways of knowing, valuing, and relating to the natural world (Menton et al., 2020). This epistemological diversity represents conceptual richness that can significantly enrich EE processes when integrated respectfully and systematically.

Eneji and Kori (2023) evaluated EE curriculum/program implementation products at the University of Calabar, Nigeria, and provided evidence of challenges in cultural adaptation. Their study of 969 EE graduates found that "the environmental education curriculum content as it stands today is not adequate" and that "the knowledge and awareness gained when studying EE did not help them sufficiently to solve their community environmental problems."

However, the study also revealed that "environmental education built their skills and prepared them well to actively participate in solving current environmental problems." This evidence suggests that cultural adaptation requires balancing local relevance with the development of transferable skills that are applicable to diverse contexts.

8.2.2 Integration of Local Worldviews and Value Systems

Effective contextual adaptation must incorporate a deep understanding of the worldviews and value systems that characterize different cultures and communities (Whitburn et al., 2020). This integration goes beyond the simple inclusion of cultural content to embrace fundamentally different ways of conceptualizing the relationship between humans and nature.

Jones (2023) reported that the new curriculum for Wales illustrates innovative approaches to integrating these areas. Her evaluation of an EE program that uses a Theory of Change framework to plan learning with 9–11-year-olds found that "combining a Theory of Change framework with a Capabilities Approach focusing on values and agency offers a more useful planning tool for curricula developing strategies where teacher agency is encouraged."

The study, based on individual and group qualitative interviews with teachers and students across six schools, revealed both the successes and challenges of this framework. The findings suggest that effective value integration requires sophisticated conceptual frameworks that connect local values to broader educational goals.

8.2.3 Socio-Environmental and Territorial Contextualization

Contextual adaptation requires a deep understanding of the specific socio-environmental characteristics of the territories where EE programs are implemented. This contextualization must account for the ecological, economic, social, and political factors that shape the relationships between human communities and natural systems.

Karim et al. (2022) highlight the significance of contextualizing climate change and EE from the perspective of science educators. Their qualitative research involving science teachers in Malaysia revealed that while the secondary school standards-based curriculum (KSSM) has seen improvements, there is a need to prioritize a student-centered approach over an exam-oriented focus.

The researchers identified "a lack of awareness of climate change and environmental action in the science textbook." They concluded that "a comprehensive curriculum, as well as empowerment among teachers and students in an environment conducive to CCEE, can generate proactive action by Malaysian students." This evidence underscores the need for effective contextualization to address both curricular content and systemic factors that influence educational implementation.

8.2.4 Linguistic and Communicative Adaptation

Effective cultural adaptation must consider the linguistic and communicative dimensions that can significantly influence the effectiveness of EE processes. This adaptation includes not only content translation but also the adaptation of communicative styles, conceptual metaphors, and culturally appropriate forms of information organization.

Effective linguistic and communicative adaptation also requires attention to educators' preparation and communicative competencies. Redman et al. (2018) examined continuing professional development in sustainability education for K-12 teachers and found that teachers who received systematic training in communicating complex environmental concepts using culturally appropriate language, metaphors, and examples demonstrated significantly stronger capacity to engage diverse student populations. Their analysis concluded that linguistic adaptation should be accompanied by sustained professional development programs that strengthen educators' ability to translate scientific environmental knowledge into accessible and culturally resonant learning experiences for their specific student communities.

8.2.5 Incorporation of Culturally Relevant Teaching Practices

Cultural adaptation requires the incorporation of culturally relevant pedagogical practices that resonate with diverse student groups. This incorporation may include traditional teaching-learning methodologies, forms of group organization, and approaches to authority and participation that reflect specific cultural norms.

Bercasio and Remolacio (2022) researched incorporating EE into the teacher-training curriculum in the Philippines and provided evidence of effective implementation strategies. Their pilot study with 2,420 third-year teacher education students found that "incorporating EE into the teacher education curriculum using DENR EE materials and EE materials made by researchers and sought out was effective in improving students' competency level in terms of environmental concepts and principles."

The study showed "substantial improvement in the EE test at a significance level of 0.05 across all ten disciplines in eight institutions." The researchers recommended "institutionalization of the incorporation of EE into the teacher training curriculum and the cross-cutting inclusion of EE in higher education programs through national policy."

8.2.6 Development of Contextually Appropriate Educational Materials

Effective contextual adaptation requires the development of educational materials that reflect the specific characteristics of the contexts in which EE programs are implemented. This development must balance local relevance, technical quality, and scientific rigor.

Salvatore and Wolbring (2022) found that the coverage of disabled people in academic literature on EE provides insights into the development of inclusive materials. Their systematic review found that "of the 73 initial sources found, only 27 contained relevant content where the content primarily engaged with disabled people as EE students but rarely with other possible roles."

The researchers concluded that there is "a need for more differentiated engagement with disabled people in EE literature." This evidence underscores the need for the development of contextually appropriate materials that systematically consider diversity and inclusion across all dimensions of human differences.

8.3 INTEGRATION OF TRADITIONAL KNOWLEDGE AND SCIENTIFIC KNOWLEDGE

8.3.1 Epistemological Frameworks for Dialogue between Different Forms of Knowledge

The effective integration of traditional and scientific knowledge in EE requires robust epistemological frameworks that can facilitate meaningful dialogue between different forms of knowing without privileging one over another. Fernández-Llamazares et al. (2021), in their comprehensive warning to humanity on threats to indigenous and local knowledge systems, argued that the erosion of indigenous knowledge represents not only a cultural loss but also a critical gap in humanity's capacity to understand and respond to environmental change. Their analysis demonstrated that indigenous and local knowledge systems contain sophisticated ecological observations accumulated over generations that are often complementary to—and in some cases more nuanced than—data derived from conventional scientific methods. The authors proposed that effective epistemological frameworks for knowledge integration must establish criteria for mutual validation that respect the internal logic and evidence standards of each knowledge tradition, create institutional spaces where knowledge holders from different traditions can engage as equals, and develop practical methodologies for synthesizing insights from diverse knowledge sources into coherent educational curricula.

8.3.2 Participatory Methodologies for Hybrid Knowledge Construction

The development of participatory methodologies that facilitate the collaborative construction of hybrid knowledge is a crucial strategy for the effective integration of traditional and scientific knowledge (Fraisl et al., 2022). These methodologies should create spaces for dialogue in which diverse forms of expertise can contribute to a richer understanding of complex socioecological systems.

Seo et al. (2020) demonstrated the construction of key competencies within an EE curriculum through the application of a modified Delphi approach in South Korea. Their research categorizes the "eight key competencies of the high school

'Environment' curriculum into three domains: intellect-oriented, personality-oriented, and relationship-oriented."

The intellect-oriented domain included "critical thinking ability, creativity, and problem-solving ability"; the personality-oriented domain included "autonomy, reflective ability, and environmental sensitivity"; and the relationship-oriented domain included "communication ability and conflict management ability." These competencies were used to review the current EE curriculum as part of the 2015 National Curriculum review.

8.3.3 Validation and Triangulation of Diverse Sources of Knowledge

The effective integration of traditional and scientific knowledge requires validation methodologies that assess the reliability and applicability of different knowledge sources without imposing a single validation criterion that may systematically favor certain forms of knowledge over others.

The use of a Delphi study to clarify the landscape and core outcomes of EE provides a methodological framework for this validation (Clark et al., 2020). Their modified Delphi study, involving 44 professionals and active leaders in North America, reached agreement on five core outcomes that focus the field of EE: "(1) environmental action and behavioral change, (2) connecting people with nature, (3) improving environmental outcomes, (4) improving social/cultural outcomes, and (5) learning environmentally relevant skills and competencies."

The researchers proposed the following statement that encapsulates the core outcomes of the field: "Environmental education works to move people to action for the tangible benefit of the environment and humanity. To realize these benefits, people must connect experientially with the environment, learn necessary skills, and understand the complex social and cultural connections between humanity and the natural environment."

8.3.4 Development of Enriching Cognitive Syntheses

Effective integration should aim to develop cognitive syntheses that can enrich both traditional and scientific knowledge through their interaction. These syntheses generate hybrid understandings that are more sophisticated and applicable than either knowledge system alone, provided that the curriculum creates genuine space for dialogue rather than merely placing both traditions side by side.

Such dialogue becomes productive when learners identify points of convergence between traditional ecological knowledge and scientific frameworks, deepening conceptual understanding while building the capacity to navigate epistemic diversity—a skill increasingly essential for addressing complex socioecological challenges.

Clark et al. (2020), through their Delphi study clarifying the landscape and core outcomes of EE, demonstrated that effective cognitive syntheses in EE emerge when programs successfully connect experiential knowledge—including traditional ecological knowledge—with formal scientific understanding. Their identification of "connecting people with nature" and "learning environmentally relevant skills and competencies" as core outcomes suggests that enriching cognitive syntheses must bridge abstract scientific knowledge with embodied, place-based understanding, creating integrated frameworks that honor multiple ways of knowing while maintaining scientific rigor. This approach aligns with Petersen et al. (2020), reporting that immersive learning technologies can facilitate deeper integration of diverse knowledge sources when designed according to evidence-based learning principles.

8.4 DIFFERENTIATION BY AGE GROUPS AND EDUCATIONAL LEVELS

8.4.1 Cognitive and Socio-Emotional Development Considerations

Effective curriculum design in EE must account for the distinctive characteristics of cognitive and socio-emotional development across age groups. This differentiation requires a sophisticated understanding of how abstract thinking, moral reasoning, and emotional processing evolve throughout human development.

Hatcher (2012) provides important insights into age differentiation by building cultural bridges with Aboriginal students and their "classmates" for transformative EE. Her analysis emphasizes that "the development of respectful relationships among all participants must precede any effective learning" and that "the development of this respect among all students results from the successful incorporation of Indigenous culture into the classroom."

The study describes how "Mi'kmaw Elder Albert Marshall describes the equal representation of knowledge from two cultural contexts as 'Seeing with Two Eyes'" and how this approach "has been very effective in the science classroom at many levels." This evidence suggests that effective age differentiation must consider not only cognitive development but also cultural and identity development, which can vary significantly across student groups, a finding consistent with the longitudinal evidence provided by Olsson et al. (2022).

8.4.2 Methodological Adaptation for Early Childhood Education

EE in early childhood requires methodologies specifically adapted to the distinctive characteristics of learning at this developmental stage, including sensory orientation, play-based learning, and the development of emotional connections with the natural world. Chawla (2020) conducted a comprehensive review of research on childhood nature connection and found that early and sustained contact with natural environments is fundamental for developing both environmental awareness and constructive hope in young children. Her analysis revealed that the most effective early childhood EE programs share common characteristics: They prioritize direct sensory experiences in nature over abstract instruction, use play as the primary vehicle for learning, build on children's innate curiosity and wonder about the natural world, and create emotionally safe spaces where children can develop positive relationships with living systems. Chawla emphasized that these early connections to nature serve as foundational experiences that shape environmental attitudes and behaviors throughout the lifespan, underscoring the critical importance of age-appropriate curriculum design in the earliest stages of education.

8.4.3 Curricular Approaches for Primary Education

EE at the primary level must balance the development of fundamental knowledge with cultivating positive emotional connections to nature. This balance requires methodologies that can make complex environmental concepts accessible to children while avoiding oversimplification, which can lead to misunderstandings.

In practice, this balance is best achieved through graduated sequencing: early grades emphasize direct sensory contact with local natural environments, while more structured ecological concepts are introduced progressively, ensuring that emotional engagement with nature precedes and sustains conceptual learning.

Olsson et al. (2022), in their longitudinal study of secondary students' action competence for sustainability, provide evidence that effective EE at earlier educational levels creates the cognitive and affective foundations upon which more complex environmental understanding is subsequently built. Their research demonstrated that students who received sustained, curriculum-integrated sustainability education from early educational stages developed progressively stronger sustainability consciousness over time, suggesting that primary education curricula should prioritize building foundational environmental knowledge and positive nature connections that serve as scaffolding for more complex learning in later stages.

8.4.4 Curriculum Differentiation for Secondary Education

EE at the secondary level can leverage more developed abstract thinking skills to address complex environmental concepts, systems analysis, and consider multiple perspectives on environmental issues. This differentiation should also consider developments in personal identity and value commitments that characterize adolescence.

This developmental stage creates a unique pedagogical window: adolescents are actively constructing their personal values and worldviews, making secondary EE particularly well positioned to engage students with competing environmental perspectives, ethical tensions, and the social dimensions of ecological decision-making.

Kopnina (2020), in her critical evaluation of education for sustainable development goals, argued that secondary education must expose students to multiple theoretical and ideological perspectives on environmental challenges rather than adopting a single conceptual framework. Her analysis emphasizes that such a transdisciplinary and multifaceted approach, informed by a combination of traditions and ideological persuasions, will offer more educational value than any single perspective could alone. This suggests that differentiation in secondary education should include structured exposure to diverse conceptual and methodological frameworks that develop students' capacity for critical environmental analysis.

Barnett et al. (2011) provide additional evidence for effective secondary-level differentiation through their urban EE program. Their study demonstrated that middle school students can effectively engage in rigorous scientific research on urban ecosystems when provided with appropriate technological tools, including geographic information systems (GIS), computational modeling, and visualization software. Students conducted scientific research on their urban ecosystems while exploring potential career options in science and technology fields. This evidence demonstrates that well-designed programs for secondary students can simultaneously develop environmental understanding, technological competencies, and career awareness.

8.4.5 Approaches to Higher Education and Professional Training

Higher-level EE should prepare students for leadership roles in environmental sustainability by developing specialized skills in research, policy analysis, organizational management, and facilitating social change (Brundiers et al., 2021).

Higher-level EE should prepare students for leadership roles in environmental sustainability by developing specialized skills in research, policy analysis, organizational management, and facilitating social change (Brundiers et al., 2021).

Veidemane (2022), in her analysis of education for sustainable development in higher education rankings, demonstrated that effective higher education programs in environmental sustainability require the integration of multidisciplinary perspectives, research-based learning, and practical engagement with real-world sustainability challenges. Her study found that institutions achieving the strongest sustainability outcomes are those that embrace interdisciplinary, learner-centered, and participatory approaches that connect academic knowledge with professional competencies. Veidemane concluded that higher education programs should give special emphasis to developing leadership capacity among diverse student populations, ensuring that sustainability professionals reflect the demographic diversity of the communities they serve.

The recommended approach "should embrace the diversity of biophysical, social, and economic environments and be closely associated with ongoing literacy, development, and functional programs" with "special emphasis given to youth, women, and policy makers." The "five basic functions for participation, which are identification, outreach, dialogue, assimilation, and feedback," are discussed.

8.4.6 Adaptation for Adult Audiences and Continuing Education

EE for adult audiences must consider the distinctive characteristics of adult learning, including a focus on practical problems, integration with life experiences, and intrinsic motivation. This adaptation requires methodologies that can connect environmental content with the professional and personal responsibilities of adult learners. Brundiers et al. (2021) identified key competencies in sustainability education that are particularly relevant for adult learners in higher education and professional contexts, including systems thinking, anticipatory competence, normative competence, and strategic competence. Their agreed-upon reference framework emphasizes that adult EE should leverage participants' existing professional knowledge and life experiences as resources for deeper environmental understanding, rather than treating them as blank slates. Furthermore, Leal Filho et al. (2020) demonstrated that universities functioning as living laboratories for sustainability provide effective models for adult EE, as they create authentic contexts where adult learners can apply environmental knowledge to real institutional challenges while developing practical skills transferable to their professional and civic lives.

8.5 CONCLUSION

8.5.1 Summary of Fundamental Principles

The analysis presented in this chapter reveals that effective curriculum design in EE requires sophisticated integration of multiple principles to navigate the complexity inherent in education processes oriented toward sustainability. The principles of comprehensiveness, contextualization, participation, progressive sequencing, formative assessment, and methodological flexibility are interdependent and must be considered holistically rather than as isolated components.

The research by Ali et al. (2025) not only provided compelling evidence that comprehensive curricular approaches can generate significant improvements in student environmental attitudes but also revealed persistent gaps between awareness and behavior that require attention to experiential and social dimensions of learning. This evidence underscores the need for effective curriculum design to simultaneously address multiple dimensions of human development.

Territorial contextualization emerges as a fundamental principle that can make abstract environmental content more relevant and meaningful for students. However, Eneji and Kori (2023) demonstrated that poorly implemented contextualization can result in programs that are locally relevant but inadequate for developing transferable skills. This tension underscores the need for curricular frameworks that balance local relevance with academic rigor and broader applicability.

8.5.2 The Imperative of Epistemological Integration

The analysis revealed that one of the most critical challenges in contemporary curriculum design is developing frameworks that effectively integrate traditional and scientific knowledge without compromising the rigor of either discipline. Research by Mbah and Ezegwu (2024) on the decolonization of EE in Africa documented the negative consequences of the systematic exclusion of indigenous and local knowledge.

This epistemological exclusion not only represents a cognitive injustice but also impoverishes the quality of EE by excluding forms of knowledge that may be particularly relevant to understanding local socioecological systems. Effective integration requires sophisticated epistemological frameworks that can recognize different criteria for validation and forms of practical application without falling into relativism, which dilutes the standards of intellectual rigor.

Clark et al. (2020) on core outcomes in EE provide evidence that it is possible to develop conceptual frameworks that integrate multiple perspectives while maintaining consistency in fundamental educational objectives. However, translating these frameworks into operational curricula requires further methodological development to address practical complexities of implementation.

8.5.3 Differentiation as a Principle of Equity

Curricular differentiation by age group and educational level is a principle of equity that recognizes that different student groups require adapted pedagogical approaches to maximize their learning potential. The research by Hatcher (2012) on EE with Aboriginal students demonstrates that effective differentiation must consider not only cognitive development but also cultural and identity characteristics that influence learning processes.

Barnett et al. (2011) demonstrated that well-differentiated urban EE programs for middle school students can effectively engage students in rigorous scientific inquiry while developing technological competencies and fostering emotional connections to urban ecosystems. This evidence suggests that effective differentiation requires a sophisticated understanding of the intersections between cognitive development, cultural context, and the characteristics of specific environmental content.

However, differentiation also carries the risk of reproducing educational inequalities if implemented in ways that limit learning opportunities for certain groups of students. Menton et al. (2020) demonstrated that EE programs that fail to account for structural inequalities risk reproducing existing disparities, underscoring the need for differentiation frameworks that provide multiple pathways to common learning objectives rather than creating separate tracks that may result in unequal access to essential knowledge.

8.5.4 Methodological and Technological Innovation

The analysis revealed that contemporary curriculum design should incorporate methodological and technological innovations to enrich environmental learning experiences without compromising the fundamental values of EE, such as a direct connection with nature and the development of emotional engagement with natural systems (Lozano et al., 2019).

Kalogiannakis and Papadakis (2017) demonstrated that the integration of mobile technologies into EE can significantly enhance educational effectiveness. This is achieved by strategically employing these technologies to connect students with pertinent environmental information within real-world contexts. However, technological integration requires care to prevent technological tools from replacing direct experiences of nature, which are fundamental for developing authentic emotional bonds.

The most promising methodological innovation appears to be one that combines the advantages of contemporary technologies with proven experiential and participatory education methodologies. This combination requires sophisticated pedagogical skills from educators and flexible curricular frameworks that can adapt to varying levels of technology availability across contexts.

8.5.5 Evaluation and Continuous Improvement

The research reviewed emphasizes that effective curriculum design must include robust formative assessment systems that provide continuous feedback on educational effectiveness and guide curriculum improvements. Polley et al. (2024) assessed threshold concepts in outdoor education and identified the importance of authentic assessment to capture complex competencies central to effective EE.

However, developing appropriate assessment systems for EE presents unique challenges because many of the most important outcomes, such as the development of environmental values, systemic thinking skills, and commitment to collective action, are difficult to measure using traditional assessment tools. Menton et al. (2020) demonstrated that EE programs that fail to account for structural inequalities risk reproducing existing disparities, underscoring the need for differentiation frameworks that provide multiple pathways to common learning objectives, rather than creating separate tracks that may result in unequal access to essential knowledge.

8.5.6 Directions for Future Development

The analysis suggests several priority directions for the future development of EE curricula. First, more research is needed on effective methodologies for integrating traditional and scientific knowledge to provide practical guidance for educators seeking to implement epistemologically inclusive approaches in their teaching (Veidemane, 2022).

Second, developing more sophisticated curriculum differentiation frameworks that account for multiple dimensions of student diversity, including cultural, linguistic, socioeconomic, and learning ability characteristics, is an important priority for promoting equity in EE (Menton et al., 2020).

Third, innovation in assessment systems that can capture complex competencies in EE while providing useful feedback for educational improvement requires sustained attention from researchers and practitioners alike (Brundiers et al., 2021).

8.5.7 Imperative for Collaborative Action

The evidence presented calls for collaborative action among researchers, educators, policymakers, and communities to develop and implement EE curricula that effectively address sustainability challenges in the 21st century (Redman et al., 2018).

This collaboration must be guided by the principles of equity, inclusion, and cultural relevance, which ensure that EE curricula serve the needs of all communities, particularly those that have been historically marginalized in formal education systems. It must also be characterized by a commitment to intellectual rigor and pedagogical effectiveness that can generate strong environmental competence in graduates.

The urgency of contemporary environmental challenges requires collaborative action on an unprecedented scale and speed. However, evidence also shows that effective curriculum development cannot be rushed; it requires careful processes of consultation, experimentation, evaluation, and refinement, which can take years to complete. Balancing urgency and care is one of the most significant challenges facing educational leadership in EE.

Finally, the success of curriculum development efforts depends critically on sustained investment in educators' capacity to effectively implement innovative curricula. This investment must include both initial training and ongoing professional development to support the long-term implementation of a high-quality curriculum.

REFERENCES

Ali, M. I., Malik, A., & Latief, M. M. (2025). Developing pro-environmental attitudes through integrated environmental education: Insights from a high school in Indonesia. *International Journal of Sustainable Development & Planning*, *20*(6), 2619–2626. https://doi.org/10.18280/ijsdp.200629

Ardoin, N. M., Bowers, A. W., & Gaillard, E. (2020). Environmental education outcomes for conservation: A systematic review. *Biological Conservation*, *241*, 108224. https://doi.org/10.1016/j.biocon.2019.108224

Barnett, M., Vaughn, M. H., Strauss, E., & Cotter, L. (2011). Urban environmental education: Leveraging technology and ecology to engage students in studying the environment. *International Research in Geographical and Environmental Education*, *20*(3), 199–214. https://doi.org/10.1080/10382046.2011.588501

Bercasio, R. R. O., & Remolacio, L. K. Q. (2022). Mainstreaming environmental education in the teacher education curriculum in the Philippines. *International Journal of Evaluation and Research in Education*, *11*(3), 1553–1564. https://doi.org/10.11591/ijere.v11i3.21748

Brundiers, K., Barth, M., Cebrián, G., Cohen, M., Diaz, L., Doucette-Remington, S., Dripps, W., Habron, G., Harré, N., Jarchow, M., Losch, K., Michel, J., Mochizuki, Y., Rieckmann, M., Parnell, R., Walker, P., & Zint, M. (2021). Key competencies in sustainability in higher education—toward an agreed-upon reference framework. *Sustainability Science*, *16*(1), 13–29. https://doi.org/10.1007/s11625-020-00838-2

Carl, J., Riley, K., Peters, J., & White, P. J. (2025). Research at the Nexus between physical education and environmental education: A narrative integrative review through a physical literacy lens. *Australian Journal of Environmental Education*, *41*(1), 4–22. https://doi.org/10.1017/aee.2025.6

Chawla, L. (2020). Childhood nature connection and constructive hope: A review of research on connecting with nature and coping with environmental loss. *People and Nature*, *2*(3), 619–642. https://doi.org/10.1002/pan3.10128

Clark, C. R., Heimlich, J. E., Ardoin, N. M., & Braus, J. (2020). Using a Delphi study to clarify the landscape and core outcomes in environmental education. *Environmental Education Research*, *26*(3), 381–399. https://doi.org/10.1080/13504622.2020.1727859

Eneji, C., & Kori, C. (2023). Products evaluation of environmental education curriculum/program implementation in the university of Calabar, Nigeria. *International Journal of Learning Teaching and Research*, *22*(4), 377–413. https://doi.org/10.26803/ijlter.22.4.22

Fernández-Llamazares, Á., Lepofsky, D., Lertzman, K., Armstrong, C. G., Brondizio, E. S., Gavin, M. C., Lyver, P. O. B., Nicholas, G. P., Pascua, P. A., Reo, N. J., Reyes-García, V., Turner, N. J., Yletyinen, J., Anderson, E. N., Balée, W., Cariño, J., David-Chavez, D. M., Dunn, C. P., Garnett, S. C., Greening, S., S. J., Kuhnlein, H., Molnár, Z., Odonne, G., Retter, G.-B., Ripple, W. J., Sáfián, L., Bahraman, A. S., Torrents-Ticó, M., & Vaughan, M. B. (2021). Scientists' warning to humanity on threats to indigenous and local knowledge systems. *Journal of Ethnobiology*, *41*(2), 144–169. https://doi.org/10.2993/0278-0771-41.2.144

Florencio da Silva, R., Torres-Rivera, A. D., Alves Pereira, V., Regis Cardoso, L., & Becerra, M. J. (2023). Critical environmental education in Latin America from a socio-environmental perspective: Identity, territory, and social innovation. *Sustainability*, *15*(12), Paper 9410. https://doi.org/10.3390/su15129410

Fraisl, D., Hager, G., Bedessem, B., Gold, M., Hsing, P.-Y., Danielsen, F., Hitchcock, C. B., Hulbert, J. M., Piera, J., Spiers, H., Thiel, M., & Haklay, M. (2022). Citizen science in environmental and ecological sciences. *Nature Reviews Methods Primers*, *2*(1), 64. https://doi.org/10.1038/s43586-022-00144-4

Hatcher, A. (2012). Building cultural bridges with Aboriginal learners and their 'classmates' for transformative environmental education. *Journal of Environmental Studies and Sciences*, *2*(4), 346–356. https://doi.org/10.1007/s13412-012-0088-6

Jones, V. (2023). Environmental education and the new curriculum for Wales: An evaluation of how a family of schools in a rural area used a theory of change approach. *Environmental Education Research*, *29*(3), 392–409. https://doi.org/10.1080/13504622.2022.2137470

Kalogiannakis, M., & Papadakis, S. (2017). Combining mobile technologies in environmental education: A Greek case study. *International Journal of Mobile Learning and Organisation*, *11*(2), 108–130. https://doi.org/10.1504/IJMLO.2017.084272

Karim, N., Othman, H., Zaini, Z.-I. I., Rosli, Y., Wahab, M. I., Al Kanta, A. M., Omar, S., & Sahani, M. (2022). Climate change and environmental education: Stance from science teachers. *Sustainability*, *14*(24), Paper 16618. https://doi.org/10.3390/su142416618

Kopnina, H. (2020). Education for the future? Critical evaluation of education for sustainable development goals. *The Journal of Environmental Education*, *51*(4), 280–291. https://doi.org/10.1080/00958964.2019.1710444

Leal Filho, W., Lange Salvia, A., Pretorius, R. W., Brandli, L. L., Manolas, E., Alves, F., Azeiteiro, U., Rogers, J., Shiel, C., & Do Paco, A. (2020). *Universities as Living Labs for Sustainable Development—Supporting the Implementation of the Sustainable Development Goals*. Springer Cham. https://doi.org/10.1007/978-3-030-15604-6

Lozano, R., Barreiro-Gen, M., Lozano, F. J., & Sammalisto, K. (2019). Teaching sustainability in European higher education institutions: Assessing the connections between competences and pedagogical approaches. *Sustainability*, *11*(6), 1602. https://doi.org/10.3390/su11061602

Mbah, M. F., & Ezegwu, C. (2024). The decolonisation of climate change and environmental education in Africa. *Sustainability*, *16*(9), Paper 3744. https://doi.org/10.3390/su16093744

Menton, M., Larrea, C., Latorre, S., Martinez-Alier, J., Peck, M., Temper, L., & Walter, M. (2020). Environmental justice and the SDGs: From synergies to gaps and contradictions. *Sustainability Science*, *15*(6), 1621–1636. https://doi.org/10.1007/s11625-020-00789-8

Olsson, D., Gericke, N., & Boeve-de Pauw, J. (2022). The effectiveness of education for sustainable development revisited: A longitudinal study on secondary students' action competence for sustainability. *Environmental Education Research*, *28*(3), 405–429. https://doi.org/10.1080/13504622.2022.2033170

Petersen, G. B., Klingenberg, S., Mayer, R. E., & Makransky, G. (2020). The virtual field trip: Investigating how to optimize immersive virtual learning in climate change education. *British Journal of Educational Technology*, *51*(6), 2099–2115. https://doi.org/10.1111/bjet.12991

Polley, S., McLeod, B., Rankin, J., Munge, B., Bovino, P., & Picknoll, D. (2024). Assessing outdoor and environmental education threshold concepts in Australian universities. *Journal of Outdoor and Environmental Education*. https://doi.org/10.1007/s42322-024-00168-2

Redman, E., Wiek, A., & Redman, A. (2018). Continuing professional development in sustainability education for K-12 teachers: Principles, programme, applications, outlook. *Journal of Education for Sustainable Development*, *12*(1), 59–80. https://doi.org/10.1177/2455133318777182

Rousell, D., & Cutter-Mackenzie-Knowles, A. (2020). A systematic review of climate change education: Giving children and young people a 'voice' and a 'hand' in redressing climate change. *Children's Geographies*, *18*(2), 191–208. https://doi.org/10.1080/14733285.2019.1614532

Salvatore, C., & Wolbring, G. (2022). Coverage of disabled people in environmental-education-focused academic literature. *Sustainability, 14*(3), Paper 1211. https://doi.org/10.3390/su14031211

Seo, E., Ryu, J., & Hwang, S. (2020). Building key competencies into an environmental education curriculum using a modified Delphi approach in South Korea. *Environmental Education Research, 26*(6), 890–914. https://doi.org/10.1080/13504622.2020.1733493

Shahat, H. A., Al-Ruwaili, H. A. S., & Aljaafari, M. Z. (2025). Environmental Education: To develop sustainability competencies related to experimentation with environmental materials among secondary school students. *International Journal of Innovative Research and Scientific Studies, 8*(2), 3152–3161. https://doi.org/10.53894/ijirss.v8i2.5958

Veidemane, A. (2022). Education for sustainable development in higher education rankings: Challenges and opportunities for developing internationally comparable indicators. *Sustainability, 14*(9), 5102. https://doi.org/10.3390/su14095102

Wamsler, C., Osberg, G., Osika, W., Herndersson, H., & Mundaca, L. (2021). Linking internal and external transformation for sustainability and climate action: Towards a new research and policy agenda. *Global Environmental Change, 71,* 102373. https://doi.org/10.1016/j.gloenvcha.2021.102373

Whitburn, J., Linklater, W., & Abrahamse, W. (2020). Meta-analysis of human connection to nature and proenvironmental behavior. *Conservation Biology, 34*(1), 180–193. https://doi.org/10.1111/cobi.13381

9 Evaluation and Impact of Educational Programs

9.1 EVALUATION FRAMEWORKS AND IMPACT INDICATORS

9.1.1 Conceptual Foundations of Evaluation in Environmental Education

Designing useful evaluation frameworks for environmental education (EE) programs requires a deep understanding of the variety and complexity of their objectives (Wamsler et al., 2021). Unlike other areas where learning is measured with more direct indicators, the aim here is to generate broader changes, not only in knowledge but also in emotions and values, behaviors, and social dynamics. Furthermore, these results are not always immediate; they may emerge over time and vary according to the context in which people live.

Hu et al. (2025) present compelling evidence regarding the efficacy of EE interventions in low- and middle-income countries through a meta-analysis encompassing 187 independent effect sizes and 34,283 participants. The results indicated "a positive and significant effect of environmental education." Specifically, participation in EE programs was associated with increases in environmental knowledge. These findings underscore the importance of developing evaluation frameworks that can capture the differentiated effects across multiple dimensions of environmental learning. Moderator analyses revealed that "results differed by intervention duration, measurement time, age, and national development level, while study design, educational level, intervention type, and gender did not show significant differences in results."

9.1.2 Development of Multidimensional Indicators

Constructing effective impact indicators requires a multidimensional approach that reflects the complexity of learning outcomes in EE (Lozano et al., 2019). Indicators must be conceptually sound, methodologically rigorous, and practically useful for different audiences, including educators, policymakers, and funding organizations.

Feucht et al. (2025) examined EE within zoological settings, highlighting the creation of specific scales designed to assess the effects of educational initiatives. Their study analyzed multiple scales to measure the impact of educational programs across nine zoos, using pre-(N = 674) and post-(N = 619) questionnaires administered to the participants in these programs. The results showed that the "adjusted instruments work well with some adaptations" and that "interest in species conservation can be measured as a general 6-item construct".

DOI: 10.1201/9781003779667-12

The study also found that environmental concern is three-dimensional, as intended, and can distinguish the effects of selfish, biospheric, and altruistic concern in zoos. This differentiation is crucial for developing indicators that can capture the multiple dimensions of environmental awareness and provide specific guidance for improving programs.

9.1.3 Theoretical Frameworks for Impact Assessment

The development of robust theoretical frameworks is essential to ensure that evaluation processes generate valid and reliable evidence of the effectiveness of EE programs (Chawla, 2020). These frameworks should integrate learning, environmental psychology, and social change theories to provide solid conceptual foundations for evaluation design.

Dong (2025) used an experimental framework to examine the impact of interactive computer-based training in nature-based EE on conservation, social, and higher- and lower-order thinking behaviors. This study compared the impact of a conservation project with and without computer technology (videos and online discussions). The results showed that the project significantly improved higher-order thinking skills, conservation behaviors, and social behaviors in both the technology and control groups.

However, no significant differences were found between the two approaches to project participation: With additional video viewing/online discussion (Group A) and without them (Group B). This finding underscores the importance of developing theoretical frameworks that can predict when and under what conditions different educational interventions are most effective.

9.1.4 Process Indicators versus Outcome Indicators

A fundamental distinction in the development of evaluation frameworks is the differentiation between process indicators, which measure the quality and fidelity of program implementation, and outcome indicators, which capture changes in participants that can be attributed to educational interventions. Both types of indicators are necessary for a comprehensive evaluation.

Gienger et al. (2024) explored the concept of responsible environmental education in the Anthropocene, focusing on the comprehension and response to young individuals' experiences of disconnection from nature, eco-anxiety, and ontological insecurity. Their narrative review of the literature found that the dominant responses of facilitating exposure to nature and encouraging environmental action risk counteracting each other and ultimately fail to address the root cause of children's and young people's experiences.

The researchers argue that emerging responses in both fields are overcoming these limitations by turning toward a reimagining humanity's relationship with nature, providing a holistic way forward. This perspective underscores the need for indicators that capture deeper paradigm shifts rather than merely superficial, behavioral changes.

9.1.5 Spatial and Temporal Scales of Assessment

Effective assessment frameworks must account for the different spatial and temporal scales at which the impacts of EE programs are manifested. These effects can range from immediate individual changes to community- and systemic-level transformations that may take years or decades to develop fully.

The availability and distribution of EE field trip programs for adolescent students in the United States provide an example of a large-scale spatial assessment. Hemby et al. (2024) provide an example of a large-scale spatial assessment through their national study, which assessed "the spatial accessibility of EE field trip programs for U.S. schools serving grades 5–8 (ages 10–14)" by compiling a comprehensive national database of 2,930 EE providers offering field trip programs and identifying 89,311 middle school locations, student populations, and relevant demographic information.

The results suggest that spatial access is highly spatially clustered, particularly in several geographic regions. Spatial access was also strongly related to partisan leanings and urbanity, with more rural, White, and Republican-leaning areas generally having significantly less spatial access to field trips. These findings underscore the importance of developing indicators that capture inequalities in access and participation, which may influence program effectiveness and equity.

9.1.6 Integrating Culturally Responsive Perspectives

The development of culturally responsive evaluation frameworks requires the recognition that different cultural groups may conceptualize, experience, and express environmental learning differently. Impact indicators must be sensitive to this diversity to avoid systematic biases that may underestimate the effectiveness of programs for specific groups.

Anderson et al. (2022) discussed a culturally responsive evaluation framework and its implementation in EE, suggesting that such frameworks enable EE organizations and evaluators to address how race, power, privilege, and inequity impact the evaluation process and the validity of its findings. The researchers argue that "implementing this framework can be resource-intensive, but it has the potential to improve evaluation processes and produce actionable results to address further issues of diversity, equity, and inclusion in the field of EE."

The proposed framework considers how issues of race, power, privilege, and inequity influence the evaluation process and emphasizes the need for organizations that support EE efforts to recognize the importance of this approach and provide adequate resources to encourage its implementation.

9.2 QUALITATIVE AND QUANTITATIVE MEASUREMENT TOOLS

9.2.1 Quantitative Assessment Methodologies

Quantitative methodologies are useful because they enable systematic and comparable measurements of changes in environmental knowledge, attitudes, and behaviors.

They rely on tools such as validated scales, controlled experiments, and advanced statistical analyses to generate robust evidence of the effectiveness of educational interventions.

van de Wetering et al. (2022) conducted a comprehensive meta-analysis of the effectiveness of EE for children and adolescents, synthesizing 50 years' worth of research. Their search identified 169 studies (512 effect sizes; 176,007 participants) conducted in 43 countries across six continents.

The results showed that EE significantly improved environmental knowledge, attitudes, intentions, and, most importantly, self-reported behaviors. However, the heterogeneity in effect sizes was substantial, and none of the moderators tested (including participant age, educational approach, and study design) explained this variance.

This meta-analysis demonstrated both the strengths and limitations of quantitative evaluation approaches. While providing clear evidence of overall effectiveness, the findings also reveal some methodological challenges for the field. The researchers identified future research priorities, including identifying effective components and approaches to EE.

9.2.2 Validated Measurement Instruments

Validated measurement instruments are essential for evaluations to provide robust, reliable, and comparable evidence of program effectiveness (Olsson et al., 2022). For these instruments to be truly useful, they must demonstrate good psychometric properties: Validity (they measure what they claim to measure), reliability (they yield consistent results), and sensitivity to change (they detect improvements or changes over time).

Ortega-Lasuen et al. (2023) conducted a study on secondary school students' knowledge of avian species and their attitudes toward conservation. The research assessed "the impact of a one-day, place-based environmental education program" implemented at the Urdaibai Bird Center. This research was based on a written questionnaire. The objective was to understand how people perceived the Urdaibai Biosphere Reserve and its wetlands, their concern for biodiversity, their knowledge of bird migration, their ability to identify different species, and their opinion on nature protection.

The results showed that "students' knowledge of Biosphere Reserves, wetlands, and bird migration is limited, and their bird identification skills are poor." However, "although they scored high on environmental attitudes, a significant number of them feel that conservation efforts are excessive and hinder economic development." These findings illustrate the importance of using multidimensional instruments that can capture both the strengths and limitations of environmental knowledge and attitudes.

9.2.3 Mixed Methodological Approaches

Mixed-method approaches combine quantitative and qualitative techniques to provide a richer and more comprehensive understanding of the impact of EE programs

(Veidemane, 2022). These approaches can capture both general patterns and specific individual experiences that can inform program improvements.

Díaz-López et al. (2023) conducted a sensitivity analysis on trends in EE within schools, employing mixed methods to examine existing initiatives, their organization across various domains, and current strategies aimed at enhancing management and implementing sustainable criteria alongside circular economy guidelines. Their analysis followed "sequential phases: (i) analysis of bibliographic records in the field of research; (ii) performance analysis and mapping of scientific production; and (iii) comparison between different policy frameworks on Environmental Education and Education for Sustainable Development."

The results showed that "Environmental Education began to be included in educational programs in the early years, taking green building practices as references" and that "over the years, it has evolved through education on consumption habits, promoting the local economy, reducing the environmental footprint, and the need to develop inclusive frameworks that address the use of technologies."

9.2.4 Direct Observation Techniques

Direct observation techniques provide unique opportunities to capture real environmental behaviors in natural contexts, thereby overcoming some limitations of self-report measures, which may be subject to social desirability bias.

Direct observation techniques also benefit from the integration of environmental and geospatial data to contextualize learning outcomes. Ardoin et al. (2020) conducted a systematic review of EE outcomes and found that the physical characteristics of learning environments—including the degree of naturalness, biodiversity, and sensory richness—can influence student-learning outcomes, although the magnitude of these effects is often modest when compared to the influence of pedagogical design and educator quality. Their review concluded that while natural settings provide important contextual support for environmental learning, the quality of the educational program itself remains the most influential factor in determining student outcomes.

These findings underscore the importance of observing and measuring multiple contextual factors simultaneously when evaluating the effectiveness of EE programs, as the quality of pedagogical design consistently emerges as more influential than the physical characteristics of the learning environment alone.

9.2.5 Participatory Evaluation Methodologies

Participatory methodologies involve participants as co-evaluators of programs, recognizing that they have unique perspectives on their own learning experiences, thereby enriching the evaluation processes.

Participatory methodologies have proven particularly valuable in community-based EE contexts, where local knowledge and perspectives are essential for understanding program impacts. Participatory methodologies have proven particularly valuable in community-based EE contexts, where local knowledge and perspectives are essential for understanding program impacts. Whitburn et al. (2020), in their meta-analysis of human connection to nature and pro-environmental behavior, found that the strength of the relationship between nature connection and behavior varies significantly across populations and contexts, with factors such as educational level, prior environmental experience, and cultural background among the most influential moderators. These findings suggest that participatory evaluation approaches—which can capture the diversity of participants' relationships with nature—are essential for understanding why programs work differently across different community contexts and demographic groups.

In addition, the authors indicated that educational level is, in general, the most influential factor in implementation and in the components of knowledge, attitudes, and practice. Finally, this approach shows how collaborative methodologies can help determine the circumstances that explain why a program works better in some groups than in others.

9.2.6 Digital Tools and Emerging Technologies

Digital tools and new technologies are expanding the ways in which EE programs can be evaluated (Petersen et al., 2020). This is because they facilitate data collection and allow for more-detailed analyses. In addition, they offer faster feedback for both students and teachers.

Repetto et al. (2024) studied 14 digital escape rooms that were designed for EE. They were based on Albert Bandura's Social Cognitive Theory. This study used a single-group quasi-experimental design. Pre- and post-test measurements were administered to 411 students aged 12–18 years. The results indicate that this strategy was highly effective in increasing self-efficacy in males. However, gender differences in initial self-efficacy require further investigation. Therefore, this case demonstrates that digital platforms can create novel educational experiences and facilitate a more rapid and organized evaluation of their effects on students.

9.2.7 Evaluating Programs in Diverse Contexts

The evaluation of EE programs is most effective when it considers the context in which it is implemented (Menton et al., 2020). It is not enough to measure whether it worked in general terms; it is also important to understand how cultural, socioeconomic, geographic, and institutional factors influence the results, as these conditions can enhance or limit the impact of any intervention.

In this regard, Uddin (2024) examined EE for sustainable development in Bangladesh. To this end, they used both qualitative and quantitative methods. Data were collected through detailed interviews, surveys administered to diverse participants, and classroom observations. The results indicate that the students lack sufficient EE and awareness. Some teachers believe that this is due to a lack of practice, low motivation, and few real actions. They also mentioned problems in institutions, inadequate policies, poor enforcement of regulations, and the influence of cultural and ancient beliefs. This study demonstrates the importance of developing assessment methods that consider the specific locations where programs are implemented.

9.3 LONG-TERM MONITORING AND SUSTAINABILITY OF RESULTS

9.3.1 Challenges of Longitudinal Monitoring

Constant monitoring of the effects of ecological teaching plans poses unique challenges regarding the conservation of subjects, handling ambiguous elements, and attributing observed changes to specific actions (Olsson et al., 2022). Even so, this type of assessment is very important for understanding the durability of educational consequences and their capacity to create stable development. For example, Zulfugarzade et al. (2024) provide an exhaustive analysis of the Russian EE research literature, providing insights into the importance of longitudinal monitoring. Their bibliometric analysis examined year-by-year distributions and references, comparisons between neighboring countries, authorship patterns, affiliations, and sources of funding.

The researchers concluded that the Russian study presented in this chapter provides a framework for future projects. Russia can be useful in this field by leveraging its experience to support the development of EE and sustainability initiatives. This finding underscores the importance of strengthening institutional capacity for conducting long-term research. This can contribute to the development of sustainable and enduring policies and programs.

9.3.2 Methodologies for Longitudinal Evaluation

Designing effective longitudinal studies requires careful planning of follow-up strategies, development of instruments appropriate for multiple measurement points, and statistical analyses that can accommodate missing data and changes in participants over time.

Designing effective longitudinal evaluation methodologies requires not only tracking behavioral and attitudinal changes over time but also developing approaches that can capture the broader societal value of EE investments. Kopnina (2020) argued that conventional approaches to evaluating EE frequently underestimate its long-term benefits because they focus narrowly on immediate learning outcomes rather than on the broader systemic contributions of EE to social transformation. Her critical analysis emphasizes that longitudinal assessment frameworks must adopt more comprehensive evaluation criteria that capture the multiple dimensions through which EE contributes to sustainability, including its effects on community resilience, environmental stewardship, and intergenerational knowledge transfer.

9.3.3 Factors Influencing the Sustainability of Results

The sustainability of EE program outcomes depends on multiple factors, including the characteristics of the original program, the implementation context, ongoing institutional support, and opportunities to apply and reinforce learning in daily life (Whitburn et al., 2020).

Thorpe et al. (2024) investigated the differential effectiveness of EE across various sociopolitical contexts within the United States. The study specifically assessed whether distinct pedagogical strategies employed during one-day EE field trip programs were associated with varying outcomes for early adolescents (aged 10–14) from diverse sociopolitical backgrounds, namely Republican majority, democratic majority, or mixed (approximately equal representation).

The findings suggest that "some traditional approaches to play, such as acting like animals or pretending to be water droplets, may feel incongruous with the social identities of students from predominantly Republican communities." This finding underscores the importance of considering the contextual factors that may influence the sustainability of educational effects in different sociopolitical environments.

9.3.4 Transferring Learning to New Contexts

A critical dimension of the sustainability of the results is the participants' ability to transfer the knowledge, skills, and attitudes developed in EE programs to new situations and contexts. This transfer is essential for educational effects to contribute to broader transformations in environmental behavior and practices.

Anderson et al. (2024) investigated the influence of perceived adult preparation and support on student attendance at a residential EE program, exploring factors that influence initial participation in programs. Their study found that "while the desire to attend residential camp did not differ by students' racial identity, Black and Latino students were less likely to attend the program."

The results indicated that "guardian support for attendance and logistical preparation were each positively related to student attendance, regardless of racial group, with Latino students reporting the lowest levels of guardian support for attendance." These findings suggest that the sustainability of outcomes may critically depend on social support factors that influence both initial participation and the continued application of learning processes.

9.3.5 Support Networks and Communities of Practice

The development of support networks and communities of practice may be crucial for sustaining the effects of EE programs beyond immediate educational experiences (Fraisl et al., 2022). These networks can provide ongoing reinforcement, opportunities for practical application, and social connections that motivate sustained actions.

Story et al. (2024) assessed barriers to participation in EE field trips in the Congaree Biosphere Reserve, and researchers conducted "semi-structured interviews with EE providers and middle school administrators." Also, they used "geospatial information systems to investigate spatial accessibility."

The results suggested that "a-spatial barriers such as curriculum constraints, capacity to plan and lead field trips, costs, and transportation influence participation." The study also found that "these a-spatial barriers are often interconnected and that access to EE field trips is disproportionately allocated to students in private education and urban settings, and that rural, poor, minority, low-achieving, and English language learners face additional obstacles."

9.3.6 Institutionalization of Educational Practices

The long-term sustainability of the effects of EE programs may require institutionalizing effective educational practices within broader systems, including formal curricula, institutional policies, and organizational cultures that support these approaches.

The long-term sustainability of EE effects requires addressing the systemic barriers that impede the institutionalization of effective practices. Redman et al. (2018) examined barriers to sustainability education institutionalization across multiple school systems and identified four interrelated categories of obstacles: Insufficient teacher training and pedagogical capacity, administrative and logistical constraints including lack of curricular time and resources, safety and liability concerns that limit outdoor and experiential learning opportunities, and attitudinal barriers among educators and administrators who may not perceive EE as a priority. Their analysis concluded that while most barriers are common across international contexts, addressing them effectively requires context-specific strategies that account for local institutional cultures, resource availability, and policy environments.

The findings revealed that "most teachers feel hampered by their lack of pedagogical knowledge to teach children" and that "while most barriers are common in global analyses, the current study also identified novel barriers." This study underscores the importance of addressing the systemic barriers that impede the sustainability of effective EE practices.

9.3.7 Assessment of Intergenerational Impacts

A particularly important dimension of long-term monitoring is the assessment of intergenerational impacts, where participants in EE programs can influence the knowledge, attitudes, and behaviors of their family members and broader communities (Ardoin et al., 2020).

Jaime et al. (2023) conducted a randomized field experiment involving fourth-grade children in Chile to assess the impact of school EE programs on fostering pro-environmental attitudes among children and their parents. Their evaluation of "the direct and indirect effects of a values-laden environmental education program on the knowledge, attitudes, and practices of children and parents regarding the consumption and disposal of plastics."

The results showed that "the educational program had a considerable and positive impact on children's knowledge, attitudes, and practices, but no effect on parents' behavior." The researchers concluded that "because parents may ultimately determine what constitutes acceptable behavior for children, promoting permanent changes in behavior will require that interventions of this type be complemented by other initiatives targeting parents."

9.4 EVIDENCE-BASED CONTINUOUS IMPROVEMENT

9.4.1 Principles of Adaptive Management in EE

Evidence-based continuous improvement requires the adoption of adaptive management principles that enable EE programs to evolve systematically, guided by evidence of implementation and outcomes (Leal Filho et al., 2020). This approach recognizes that effective EE is an iterative process that requires continuous adjustment based on emerging learning.

In identifying areas and approaches for improving assessment processes in EE in the United States, Anderson et al. (2022) used an online survey of US professionals to explore "professionals' satisfaction with their assessment processes and the drivers of this satisfaction." The results indicated "low satisfaction with evaluation processes in general, but greater satisfaction among organizations involved in systematic formal evaluation, particularly processes focused on adaptive management and program improvement."

The study found that organizations had higher satisfaction "when compared to evaluations focused on meeting external accountability requirements or using informal evaluation processes based on non-systematic observation and reflection." The results also highlighted "current gaps in using evaluation for adaptive management and to promote diversity, equity, and inclusion efforts in the field."

9.4.2 Feedback Systems and Programmatic Adjustment

Developing effective feedback systems requires mechanisms that capture relevant information from multiple sources, process it in a timely and meaningful manner, and translate it into concrete adjustments to improve program effectiveness.

Bird et al. (2022) offer an analysis of program evaluation practices from the perspective of the professional evaluator. Their ten-year meta-evaluation (2008–2018) of data collected from a suite of place-based learning programs delivered by a charity in Ireland yielded insights into the impact of time constraints on evaluation, evaluation as part of organizational culture, the strengths and weaknesses of evaluations led by professional evaluators, and the opportunities that meta-evaluation offers for organizational change.

The findings show that meta-evaluation, or the assessment of the quality of an evaluation, is highly useful for experts and researchers. This study provides clear guidance for designing and improving future evaluations. In summary, this approach demonstrates that feedback helps in many ways; it enables rapid program changes (e.g., what to improve and what to fix). It also provides data for larger, enduring organizational changes over time.

9.4.3 Integration of Multiple Stakeholder Perspectives

Effective continuous improvement requires the integration of multiple stakeholder perspectives, including those of participants, educators, administrators, funders, and community members. Each group can provide unique insights into different aspects of the programs, informing improvement strategies.

Effective continuous improvement also requires integrating the perspectives of educators regarding which pedagogical techniques are most appropriate for different developmental stages. Rousell and Cutter-Mackenzie-Knowles (2020) conducted a systematic review examining how EE programs engage children and young people in civic participation and advocacy around climate change. Their analysis found that educators consistently rated participatory and action-oriented techniques as highly appropriate but acknowledged that many such techniques—particularly those related to civic engagement, advocacy, and behavioral change for diversity, equity, and inclusion—were underutilized in practice. The authors identified a significant gap between what educators consider appropriate and what they actually implement, suggesting that continuous improvement efforts should focus on building educators' confidence and institutional support for employing more ambitious participatory methodologies.

The results showed that "educators rated many techniques as less appropriate and were less likely to use them with younger age groups." The study also conducted "an appropriateness/use analysis to identify techniques that were considered highly appropriate but not used as frequently" and found that "techniques related to civic engagement and advocacy for diversity, equity, and inclusion appeared to be underutilized."

9.4.4 Documentation and Systematization of Experiences

The systematic documentation of programmatic experiences is fundamental to building organizational knowledge and developing best practices that can be shared across different contexts (Sonetti et al., 2020). This documentation should capture both successes and failures to provide comprehensive learning outcomes.

Kabassi (2021) conducted a study comparing multicriteria decision-making models to evaluate EE programs. The research examined the comparative performance of the simple additive weighting (SAW), weighted product method (WPM), technique for order of preference by similarity to ideal solution (TOPSIS), and Preference Ranking Organization Method for Enrichment Evaluations II (PROMETHEE II) models in assessing and selecting the most suitable EE program. The main objective of the study was to "present the different steps of the comparative analysis of multi-criteria decision-making models and draw conclusions about the suitability and robustness of the SAW, WPM, TOPSIS, and PROMETHEE II models in evaluating environmental education programs."

This type of comparative analysis can inform the development of decision-making tools that help organizations select and adapt programmatic approaches based on systematic evidence of their relative effectiveness across contexts.

9.4.5 Organizational Capacity Building

Continuous improvement requires the development of organizational capacities that enable institutions to implement systematic cycles of planning, implementation, evaluation, and adjustment. These capacities include technical competencies, information management systems, and organizational cultures that value learning and adaptation.

Building organizational capacity for continuous improvement also requires investing in the professional development of educators who will implement and evaluate EE programs. Building organizational capacity for continuous improvement also requires investing in the professional development of educators who will implement and evaluate EE programs. Brundiers et al. (2021), in their agreed-upon reference framework for key competencies in sustainability education, identified systems thinking, anticipatory competence, normative competence, and strategic action competence as essential capabilities that educators need to implement effective EE programs. Their framework, developed through extensive consultation with sustainability education experts across multiple institutions, emphasizes that building institutional capacity requires sustained investment in developing these competencies among educators through a combination of formal training, mentoring, and collaborative learning opportunities that connect theoretical understanding with practical application in diverse educational contexts.

The authors concluded that the development of independent EE courses within teacher preparation programs, complemented by ongoing professional development, is essential for building the institutional capacity needed to sustain effective EE practices. The findings suggested "the importance of independent EE courses and student-centered activities to develop students' behavior toward the environment." This research illustrates how capacity building can occur through educator training, enabling educators to implement improved practices in their work contexts.

9.4.6 Innovation and Programmatic Experimentation

Continuous improvement must include space for innovation and programmatic experimentation, enabling organizations to explore new approaches, methodologies, and technologies to enhance the effectiveness of EE programs.

Stern et al. (2025) conducted a mixed-methods study to explore principles for enhancing EE field trips for early adolescents in the United States. The study involved the observation of "87 EE-focused school field trips in the U.S.," followed by immediate surveys of the young participants. The researchers used "student surveys to quantitatively distinguish the top quartile of programs from the bottom quartile." Then they examined "in-depth qualitative field notes on each program in the top and bottom quartiles."

Their analysis revealed "six key principles that distinguished programs with the most positive student outcomes from those with the least positive outcomes: (1) Ensuring that basic needs (physiological and safety) are met; (2) ensuring instructional clarity; (3) providing emotional support for students; (4) emphasizing the novelty of activities and environment; (5) maintaining or restoring student attention; and (6) providing a cohesive, thematic experience with clear purpose."

This study demonstrates how systematic approaches to documenting and analyzing program practices can generate evidence-based principles that can inform future innovations in program design and implementation.

9.5 CONCLUSION

9.5.1 Synthesis of Challenges and Opportunities in Evaluation

The study presented in this section indicates that analyzing ecological education projects poses unique and complex challenges, requiring advanced analytical methods and comprehensive theoretical frameworks. Problems include the need to account for varied effects that may manifest across temporal and spatial dimensions, the difficulty in attributing observed changes to specific activities, and the importance of developing indicators that are culturally adapted and appropriately contextualized.

van de Wetering et al. (2022) identified positive and robust effects across various contexts and methodologies. Nevertheless, the substantial variability in effect sizes suggests a need for further investigation into the conditions that account for the differential effectiveness of programs across contexts.

The research also revealed the critical importance of considering the contextual factors that can influence program effectiveness. Furthermore, Thorpe et al. (2024) found differences in the effectiveness of play-based pedagogies in different sociopolitical contexts, illustrating how cultural and political factors can mediate the impact of educational interventions. This evidence highlights the need to develop evaluation approaches that are sensitive to the diverse contexts in which EE programs are implemented.

9.5.2 Integration of Quantitative and Qualitative Approaches

A central conclusion is that the most robust evaluations in EE typically employ mixed approaches in which quantitative and qualitative methods complement each other. On the one hand, quantitative methods allow for a rigorous estimation of the magnitude and significance of program effects. In contrast, qualitative methods help to understand how and why changes occur, revealing processes, experiences, and contextual factors that may explain differences in effectiveness.

EE in zoos demonstrates that validated measurement tools substantially strengthen evaluations by enabling rigorous assessments of the dimensions of environmental learning. Concurrently, the meta-evaluation conducted by Bird et al. (2022) underscores the significance of qualitative methodologies in elucidating the functioning of an organization's internal processes and in facilitating the enhancement of programs from within.

For this combination to work, conceptual frameworks are needed that integrate different types of evidence and analysis strategies that allow for triangulation, that is, to contrast and connect findings from various sources. This is key when formulating recommendations for improvement that are not only well supported by data but also viable and useful for those who implement the programs in practice.

9.5.3 Sustainability and Transfer of Impacts

Jaime et al. (2023) examined the intergenerational effects of educational programs, highlighting both the potential and the constraints of learning transfer across generations. While programs can have significant effects on direct participants, their transfer to other family members and communities may require additional intervention strategies.

The sustainability of the results depends on multiple factors, including the design of the original programs, the presence of ongoing support systems, and opportunities to apply and reinforce learning in everyday contexts. The findings of Anderson et al. (2024) on factors influencing student participation underscore the importance of addressing systemic barriers that can impede both initial participation and the ongoing application of learning.

9.5.4 Equity and Accessibility in Assessment

The analysis revealed significant concerns regarding equity and accessibility in both program participation and assessment processes. Hemby et al. (2024) documented significant geographic disparities in access to EE programs, which can affect who can benefit from them. Similarly, Anderson et al. (2022) emphasized the need to address how issues of race, power, and privilege can influence assessment processes in their research on culturally responsive assessment frameworks.

These concerns underscore the need to develop evaluation approaches that not only measure effectiveness but also examine equity in access, participation, and benefits. Future evaluation frameworks should include specific indicators that capture disparities and provide evidence to inform the addressing of systemic inequalities in EE provision and assessment.

9.5.5 Technological Innovation in Evaluation

The analysis identified significant opportunities for technological innovation in the evaluation of EE programs. Repetto et al. (2024) demonstrated how emerging technologies can provide innovative platforms for both program implementation and evaluation in their research on digital escape rooms. Digital technologies can facilitate more efficient data collection, sophisticated analysis, and immediate feedback for participants and educators.

However, the adoption of innovative technologies must be accompanied by careful consideration of accessibility, privacy, and equity issues. Technological tools must be designed for effective use by organizations with varying levels of technical capacity and financial resources.

9.5.6 Future Directions for Research and Practice

Based on the evidence reviewed, several future directions emerge as priorities for developing EE evaluation. First, more research is needed on the causal mechanisms linking specific educational interventions to learning outcomes, particularly with respect to the contextual factors that may moderate these effects. Ardoin et al. (2020) identified this as a critical gap in their systematic review, noting that while evidence consistently supports the overall effectiveness of EE, understanding the specific mechanisms through which different program components contribute to outcomes remains a priority for advancing the field.

Second, developing longitudinal evaluation frameworks that capture long-term changes and impacts is a priority. These frameworks should include strategies for maintaining participant engagement over extended periods and methods for distinguishing programmatic effects from other factors that may influence participant's development.

Third, developing collaborative evaluation systems involving multiple stakeholders in the design, implementation, and interpretation of evaluations can enhance the quality and usefulness of the generated evidence. These systems must balance methodological rigor with the authentic participation of different stakeholder groups.

9.5.7 Imperative for Coordinated Action

The evidence presented calls for coordinated action among researchers, educators, policymakers, and funding organizations to strengthen the evaluation systems in EE (Kopnina, 2020). This coordination should focus on developing quality standards for evaluation, building organizational capacities to implement effective evaluations, and creating mechanisms for sharing and using evaluation evidence.

The urgency of contemporary environmental challenges requires coordinated action. However, the complexity of effective EE systems also demands care and rigor in developing assessment approaches. Balancing urgency and quality are a fundamental challenge for leadership in EE.

Finally, the success of assessment improvement depends critically on organizational cultures that value evidence, learning, and continuous improvement. Building these cultures requires sustained investment in capacity building, technical support systems, and incentive frameworks that reward the effective use of assessment evidence for program improvement.

REFERENCES

Anderson, K. C., Stern, M. J., & Powell, R. B. (2022). Identifying areas and approaches for improving evaluation processes in environmental education in the United States of America. *The Journal of Environmental Education*, *53*(5), 290–303. https://doi.org/10.1080/00958964.2022.2122918

Anderson, K. C., Stern, M. J., & Powell, R. B. (2024). Investigating the influence of preparation and perceived adult support on student attendance to a residential environmental education program. *Environmental Education Research*, *30*(2), 251–264. https://doi.org/10.1080/13504622.2023.2247586

Ardoin, N. M., Bowers, A. W., & Gaillard, E. (2020). Environmental education outcomes for conservation: A systematic review. *Biological Conservation*, *241*, 108224. https://doi.org/10.1016/j.biocon.2019.108224

Bird, Á., Fahy, F., & Reilly, K. (2022). Making evaluation work for the practitioner evaluator: Experience from the field of environmental education. *Environmental Education Research*, *28*(5), 715–734. https://doi.org/10.1080/13504622.2022.2036324

Brundiers, K., Barth, M., Cebrián, G., Cohen, M., Diaz, L., Doucette-Remington, S., Dripps, W., Habron, G., Harré, N., Jarchow, M., Losch, K., Michel, J., Mochizuki, Y., Rieckmann, M., Parnell, R., Walker, P., & Zint, M. (2021). Key competencies in sustainability in higher education—toward an agreed-upon reference framework. *Sustainability Science*, *16*(1), 13–29. https://doi.org/10.1007/s11625-020-00838-2

Chawla, L. (2020). Childhood nature connection and constructive hope: A review of research on connecting with nature and coping with environmental loss. *People and Nature, 2*(3), 619–642. https://doi.org/10.1002/pan3.10128

Díaz-López, C., Serrano-Jiménez, A., Chacartegui, R., Becerra-Villanueva, J. A., Molina-Huelva, M., & Barrios-Padura, Á. (2023). Sensitivity analysis of trends in environmental education in schools and its implications in the built environment. *Environmental Development, 45*, 100795. https://doi.org/10.1016/j.envdev.2022.100795

Dong, Q. (2025). Computer-based interactive training in nature-based environmental education: Impact on conservation behaviours, social behaviours, and lower and higher order thinking skills. *Thinking Skills and Creativity, 57*, 101825. https://doi.org/10.1016/j.tsc.2025.101825

Feucht, V., Dierkes, P. W., Haeser-Kalthoff, A., Kästner, A., Mager, C., Reichhardt, S., Schröder, L., Spengler, T., Steinert, I., Ugolini, P., Waked, F., & Kleespies, M. W. (2025). Environmental education in zoos: Analysis of different scales to measure the impact of educational programs. *Environmental Education Research, 31*(4), 776–793. https://doi.org/10.1080/13504622.2024.2386631

Fraisl, D., Hager, G., Bedessem, B., Gold, M., Hsing, P.-Y., Danielsen, F., Hitchcock, C. B., Hulbert, J. M., Piera, J., Spiers, H., Thiel, M., & Haklay, M. (2022). Citizen science in environmental and ecological sciences. *Nature Reviews Methods Primers, 2*(1), 64. https://doi.org/10.1038/s43586-022-00144-4

Gienger, A., Nursey-Bray, M., Rodger, D., Szorenyi, A., Weinstein, P., Hanson-Easey, S., Fordham, D., Lemieux, D., Hill, C., & Yoneyama, S. (2024). Responsible environmental education in the Anthropocene: Understanding and responding to young people's experiences of nature disconnection, eco-anxiety and ontological insecurity. *Environmental Education Research, 30*(9), 1619–1649. https://doi.org/10.1080/13504622.2024.2367022

Hemby, T. L., Powell, R. B., & Stern, M. J. (2024). Availability and distribution of environmental education field trip programs for adolescent students in the US: A national study of spatial accessibility. *Environmental Education Research, 30*(2), 214–234. https://doi.org/10.1080/13504622.2023.2237706

Hu, B., Wang, C., Xia, Z., Lu, C., & Ju, X. (2025). Environmental education in low-income and middle-income countries: A systematic review and meta-analysis. *Journal of Environmental Psychology, 104*, 102613. https://doi.org/10.1016/j.jenvp.2025.102613

Jaime, M., Salazar, C., Alpizar, F., & Carlsson, F. (2023). Can school environmental education programs make children and parents more pro-environmental? *Journal of Development Economics, 161*, 103032. https://doi.org/10.1016/j.jdeveco.2022.103032

Kabassi, K. (2021). Comparing multi-criteria decision making models for evaluating environmental education programs. *Sustainability, 13*(20), 11220. https://doi.org/10.3390/su132011220

Kopnina, H. (2020). Education for the future? Critical evaluation of education for sustainable development goals. *The Journal of Environmental Education, 51*(4), 280–291. https://doi.org/10.1080/00958964.2019.1710444

Leal Filho, W., Eustachio, J. H., Caldana, A. C., Will, M., Lange Salvia, A., Rampasso, I. S., Anholon, R., Platje, J., & Kovaleva, M. (2020). Sustainability leadership in higher education institutions: An overview of challenges. *Sustainability, 12*(9), 3761. https://doi.org/10.3390/su12093761

Lozano, R., Barreiro-Gen, M., Lozano, F. J., & Sammalisto, K. (2019). Teaching sustainability in European higher education institutions: Assessing the connections between competences and pedagogical approaches. *Sustainability, 11*(6), 1602. https://doi.org/10.3390/su11061602

Menton, M., Larrea, C., Latorre, S., Martinez-Alier, J., Peck, M., Temper, L., & Walter, M. (2020). Environmental justice and the SDGs: From synergies to gaps and contradictions. *Sustainability Science*, *15*(6), 1621–1636. https://doi.org/10.1007/s11625-020-00789-8

Olsson, D., Gericke, N., & Boeve-de Pauw, J. (2022). The effectiveness of education for sustainable development revisited: A longitudinal study on secondary students' action competence for sustainability. *Environmental Education Research*, *28*(3), 405–429. https://doi.org/10.1080/13504622.2022.2033170

Ortega-Lasuen, U., Pedrera, O., Telletxea, E., Barrutia, O., & Díez, J. R. (2023). Secondary students' knowledge on birds and attitudes towards conservation: Evaluation of an environmental education program. *International Journal of Environmental Research and Public Health*, *20*(10), 5769. https://doi.org/10.3390/ijerph20105769

Petersen, G. B., Klingenberg, S., Mayer, R. E., & Makransky, G. (2020). The virtual field trip: Investigating how to optimize immersive virtual learning in climate change education. *British Journal of Educational Technology*, *51*(6), 2099–2115. https://doi.org/10.1111/bjet.12991

Redman, E., Wiek, A., & Redman, A. (2018). Continuing professional development in sustainability education for K-12 teachers: Principles, programme, applications, outlook. *Journal of Education for Sustainable Development*, *12*(1), 59–80. https://doi.org/10.1177/2455133318777182

Repetto, M., Bianco Prevot, A., Brizio, A., Boldi, A., Talarico, M., Stanchi, S., Palma, D., & Acquadro Maran, D. (2024). Digital escape rooms: A resource for environmental education. *Sustainability*, *16*(19). https://doi.org/10.3390/su16198525

Rousell, D., & Cutter-Mackenzie-Knowles, A. (2020). A systematic review of climate change education: Giving children and young people a 'voice' and a 'hand' in redressing climate change. *Children's Geographies*, *18*(2), 191–208. https://doi.org/10.1080/14733285.2019.1614532

Sonetti, G., Barioglio, C., & Campobenedetto, D. (2020). Education for sustainability in practice: A review of current strategies within Italian universities. *Sustainability*, *12*(13), 5246. https://doi.org/10.3390/su12135246

Stern, M. J., Powell, R. B., Frensley, B. T., Anderson, K. C., Kiewra, L., Pownall, M., Schmitt, T., & Thorpe, E. G. (2025). Principles for better environmental education field trips for early adolescent youth in the United States: An empirical study. *Environmental Education Research*, 1–24. https://doi.org/10.1080/13504622.2025.2502575

Story, T., Powell, R. B., Baldwin, E., Baldwin, R. F., & Dertien, J. S. (2024). Assessing barriers to participation in environmental education field trips in the Congaree Biosphere Reserve. *Environmental Education Research*, *30*(6), 964–986. https://doi.org/10.1080/13504622.2023.2246697

Thorpe, E. G., Stern, M. J., Powell, R. B., & Hemby, T. L. (2024). Does environmental education work differently across sociopolitical contexts in the United States? Part II. Examining pedagogy in school field trip programs for early adolescent youth across political contexts. *Environmental Education Research*, *30*(5), 753–774. https://doi.org/10.1080/13504622.2023.2295781

Uddin, M. K. (2024). Environmental education for sustainable development in Bangladesh and its challenges. *Sustainable Development*, *32*(1), 1137–1151. https://doi.org/10.1002/sd.2728

van de Wetering, J., Leijten, P., Spitzer, J., & Thomaes, S. (2022). Does environmental education benefit environmental outcomes in children and adolescents? A meta-analysis. *Journal of Environmental Psychology*, *81*, 101782. https://doi.org/10.1016/j.jenvp.2022.101782

Veidemane, A. (2022). Education for sustainable development in higher education rankings: Challenges and opportunities for developing internationally comparable indicators. *Sustainability, 14*(9), 5102. https://doi.org/10.3390/su14095102

Wamsler, C., Osberg, G., Osika, W., Herndersson, H., & Mundaca, L. (2021). Linking internal and external transformation for sustainability and climate action: Towards a new research and policy agenda. *Global Environmental Change, 71*, 102373. https://doi.org/10.1016/j.gloenvcha.2021.102373

Whitburn, J., Linklater, W., & Abrahamse, W. (2020). Meta-analysis of human connection to nature and proenvironmental behavior. *Conservation Biology, 34*(1), 180–193. https://doi.org/10.1111/cobi.13381

Zulfugarzade, T. E., Chirkina, S. E., Kondrashev, S. V., Chernega, T., Luzin, A., & Tararina, L. I. (2024). A comprehensive bibliographic analysis of environmental education research in Russia. *Eurasia Journal of Mathematics, Science and Technology Education, 20*(4), em2430. https://doi.org/10.29333/ejmste/14425

Section IV

Implementation and Future of Environmental Education

10 Effective Implementation Strategies

10.1 PROGRAM DESIGN: FROM THEORY TO PRACTICE

10.1.1 Fundamental Principles of Program Design

Effective environmental education (EE) program design requires integrating sound theoretical principles with a deep understanding of the practical realities of implementation contexts (Redman et al., 2018). This integration must begin with a rigorous diagnostic phase to identify both specific educational needs and the contextual opportunities and constraints that will influence the viability and effectiveness of different programmatic approaches to teacher training.

Gilleran Stephens et al. (2025) highlight the significance of employing Universal Design for Learning (UDL) as a framework for redesigning EE programs, emphasizing the need to incorporate inclusive principles into program design. Their study found that "environmental education remains marginalized in Irish primary schools despite its formal inclusion in the curriculum" and that UDL-based programs can overcome barriers related to "curricular pressures, time constraints, conceptual misunderstandings, and logistical concerns."

The proposed six-step UDL redesign process, which includes planning, designing, implementing, evaluating, testing, and refining, provides a systematic framework that can be applied to other EE programs and science, technology, engineering, and mathematics (STEM) outreach initiatives. Findings indicated that "hands-on, experiential learning significantly improved both teacher and student engagement."

10.1.2 Conceptual Frameworks for Theory-Practice Integration

Transforming theoretical frameworks into operational programs requires conceptual frameworks that can articulate the intervention logic, causal assumptions, and theories of change that guide program design. These frameworks should explicitly articulate the connections among specific activities, intermediate outcomes, and long-term impacts, providing conceptual maps to guide both implementation and evaluation.

Wu et al. (2025) present a comprehensive conceptual framework that elucidates the mechanisms by which EE influences pro-environmental behavior. Their research, conducted in Wuyishan National Park, employs the stimulus-organism-response (SOR) paradigm to develop a model that delineates the relationship between educational perception and pro-environmental behavior.

The findings revealed that "pro-environmental behavior is directly and significantly promoted by the perception of environmental education, place attachment, and environmental attitude" and that "place attachment and environmental attitude play an intermediary role between the perception of environmental education and

DOI: 10.1201/9781003779667-14

pro-environmental behavior." Furthermore, "the quality of the external environment positively influences the correlation between the perception of environmental education and place attachment."

10.1.3 Methodologies for Programmatic Contextualization

Effective contextualization requires methodologies that capture the specific characteristics of implementation contexts and translate this understanding into appropriate program adaptations. Ardoin et al. (2020) conducted a systematic review of EE outcomes and found that programs incorporating empathy-building strategies, reflective thinking components, and culturally responsive pedagogies consistently demonstrated stronger outcomes across cognitive, affective, and behavioral dimensions. Their analysis revealed that contextualized programs that integrate emotional engagement with critical reflection produce more durable behavioral changes than information-only approaches, particularly when designed to address the specific environmental challenges and cultural realities of the target population.

10.1.4 Development of Specific Theories of Change

Theories of change provide logical frameworks that articulate how specific programmatic activities can generate the desired changes in knowledge, attitudes, behaviors, and environmental conditions. These theories should be specific enough to guide design and implementation decisions, yet flexible enough to accommodate adaptations in response to emerging insights from the study.

Theories of change in EE must account for how educational experiences reshape learners' perceptions of and relationships with their environments. Chawla (2020) conducted a comprehensive review of research on childhood nature connection and found that children's environmental representations are dynamic and responsive to structured educational interventions. Her analysis revealed that young children who participate in place-based EE programs develop more complex and nature-oriented representations of their local environments, moving beyond purely recreational framings toward understanding ecological relationships and developing a sense of environmental stewardship. These findings suggest that specific theories of change for early childhood EE should explicitly model the progression from initial perceptions to more sophisticated environmental understanding.

10.1.5 Adaptation Mechanisms and Continuous Learning

Effective program design must include mechanisms that enable continuous adaptation in response to emerging evidence of effectiveness, changes in implementation contexts, and an evolving understanding of environmental issues. These mechanisms require monitoring systems that provide timely feedback and organizational processes that translate this into programmatic adjustments.

Gal (2024) provided insights into adaptation mechanisms by examining the strengths, weaknesses, opportunities, and threats (SWOT) of a long-term outdoor EE program in Israel. Their SWOT analysis of a program implemented continuously since 1996 found that strengths and opportunities appear to outweigh threats and weaknesses, supporting the program's long-term success.

The study revealed that long-term outdoor EE programs that achieve EE goals over time can have significant positive impacts, both environmental and on the communities involved, and remain sustainable when appropriate attention is paid to relationships with experts and funding organizations.

10.2 IMPLEMENTATION IN DIVERSE CONTEXTS (URBAN, RURAL, INDIGENOUS)

10.2.1 Specific Considerations for Urban Contexts

Implementing EE programs in urban contexts presents unique opportunities and challenges, including high population density, cultural diversity, limited access to natural spaces, and complex urban socioecological systems. Urban programs must navigate these characteristics to create educational experiences that connect participants with natural processes often hidden or altered in urban environments.

Chávez Santos et al. (2024) examine the transformative impact of EE on safe water consumption within a low-income neighborhood, highlighting specific strategies applicable to marginal urban settings. Their study evaluated the impact of an educational intervention in a community without access to safe drinking water supplied by shallow wells using a pre-experimental quantitative design with three stages: Diagnosis, educational intervention, and evaluation.

The results showed that educational intervention significantly improved knowledge and habits related to safe water consumption ($p < 0.001$), and these changes persisted after the intervention, with no significant differences by gender or educational level. The study concluded that short-term EE programs can positively change the behavior of vulnerable populations without access to safe drinking water.

10.2.2 Strategies for Implementation in Rural Contexts

Rural contexts have distinctive characteristics, including greater proximity to natural systems, frequent economic dependence on natural resources, and social structures that can facilitate community-learning processes. However, these contexts may also face limitations in access to educational resources, technological connectivity, and specialized institutional capacities.

Rural contexts benefit particularly from integrating traditional ecological knowledge with formal EE approaches. Fernández-Llamazares et al. (2021), in their comprehensive warning to humanity on threats to indigenous and local knowledge systems, emphasized that traditional ecological knowledge encompasses a complex understanding of the interplay between ecological processes, resource management, and community well-being. Their analysis demonstrated that the legal recognition and protection of traditional knowledge, combined with its systematic integration into formal education systems, can significantly strengthen both EE outcomes and community-based resource management practices. The authors concluded that rural implementation strategies should actively document, protect, and incorporate local ecological knowledge through community programs and educational initiatives that bridge traditional and scientific knowledge systems.

10.2.3 Specific Approaches for Indigenous Communities

The implementation of indigenous contexts requires conceptual and methodological frameworks that can honor and integrate traditional knowledge systems, community governance structures, and specific relationships with territories and natural resources. These approaches must overcome the legacies of educational colonization to create culturally appropriate and empowering learning spaces.

Dodson and Miru (2021), in their study of self-determination through indigenous EE in New Zealand, demonstrate innovative approaches in indigenous contexts. Their study of the Ngā Waihotanga Iho estuary monitoring kit, as a central component of a Māori-centered educational project, found that the kit was designed to support kaitiakitanga (environmental stewardship) and indigenous-led EE.

The study demonstrated that approaches such as this can be powerful vehicles for indigenous self-determination, as Māori actively undertake tribal development and environmental stewardship and strengthen the place of indigenous knowledge, priorities, and approaches within an evolving "postcolonial" education system.

10.2.4 Cultural and Linguistic Adaptation

Effective cultural and linguistic adaptation requires more than simple translation; it demands integrating local knowledge, cultural practices, and community experiences into the educational framework itself. Menton et al. (2020) examined the tensions between global sustainability agendas and local community realities, finding that EE programs are most culturally authentic when they explicitly document, elevate, and celebrate local ecological knowledge rather than imposing externally developed curricula. Their analysis of environmental justice dimensions across diverse cultural contexts demonstrated that meaningful cultural adaptation must incorporate local experiences, histories, and knowledge systems to create programs that resonate with community values and address locally relevant environmental challenges.

10.2.5 Contextual Factors Influencing Effectiveness

Contextual factors significantly influence the effectiveness of EE programs, including socioeconomic conditions, prior environmental experience, and the integration of EE with local economic activities such as tourism. Whitburn et al. (2020) conducted a meta-analysis examining contextual moderators of the relationship between nature connection and pro-environmental behavior. Their analysis found that EE plays a significant role in strengthening the relationship between environmental knowledge and sustainable practices and that this moderating effect is particularly pronounced in communities where economic livelihoods are directly connected to natural resource management, such as tourism-dependent regions. These findings suggest that implementation strategies should carefully assess and leverage the specific contextual factors that can amplify or attenuate the effects of EE interventions.

10.3 SCALING AND REPLICABILITY OF SUCCESSFUL INITIATIVES

10.3.1 Principles for Effective Scaling

Scaling up successful EE initiatives requires conceptual frameworks that identify essential elements to be maintained during expansion and components that can and should be adapted to new contexts (Olsson et al., 2022). This scaling up must balance fidelity to fundamental principles with the flexibility for contextual adaptation.

Research on Recycling Organization through Education for Sustainability (ROSE) investigates the mediating effects of waste minimization behavior, offering valuable insights into scalability in higher education (Sobaih et al., 2024). The study, which involved 597 students at King Faisal University in Saudi Arabia, demonstrated that EE programs have a significant and positive impact on environmental performance.

The results showed a significant positive relationship between EE programs and waste minimization. Waste minimization behavior partially mediates the relationship between EE programs and environmental performance. This model can provide a framework for scaling up other higher education institutions.

10.3.2 Transfer and Adaptation Models

The development of effective transfer models requires frameworks that distinguish between core elements that must be maintained and peripheral components that can be modified in response to specific contextual characteristics of the target environment. These models should provide practical guidance for organizations seeking to adapt successful initiatives to their contexts.

Ostrem and Hvenegaard (2025) examine the interagency collaboration for environmental education and offer insights into transfer models. Their research, conducted in the Beaver Hills Biosphere in Alberta, Canada, uses 23 semi-structured interviews to elucidate current mechanisms of collaboration and their associated benefits and disadvantages. Additionally, the study identifies facilitators and barriers and provides recommendations for the future within the framework of collective impact theory.

The findings highlighted "the principles and importance of interagency collaboration across all stages of delivering environmental education, from creation and leadership to delivery and evaluation." This collaborative model can be transferred to other biosphere regions and conservation contexts.

10.3.3 Strategies for Institutionalization

The sustainability of scaled-up initiatives requires institutionalization strategies that embed successful programs within permanent organizational structures, policies, and cultures capable of sustaining them over extended periods.

The institutionalization of EE within organizational contexts requires both leadership commitment and systematic knowledge-sharing mechanisms. Lozano et al. (2019) examined the connections between sustainability competences and pedagogical approaches across European higher education institutions and found that successful institutionalization depends on the sustained engagement of organizational leaders who champion EE, combined with robust systems for sharing environmental knowledge across departments and organizational levels. Their analysis revealed that institutions where leadership actively promoted EE and established formal knowledge-sharing activities demonstrated significantly stronger integration of sustainability into both educational and operational practices, suggesting that institutionalization requires parallel investment in leadership development and organizational learning infrastructure.

10.3.4 Networks and Learning Platforms

Effective scaling often requires developing networks and platforms that facilitate the exchange of experiences, resources, and knowledge among organizations implementing similar initiatives. These networks can provide technical support, approach validation, and opportunities for collaborative innovation.

The importance of networks for scaling EE is illustrated by the challenges identified in cross-sectoral implementation efforts. Kola-Olusanya et al. (2023), in their study of the role of EE in water pollution prevention and conservation in Nigeria, found that despite high levels of awareness among 766 respondents, attitudes toward prevention remained poor. Their recommendation that the Federal Ministries of Water Resources and Education should collaborate to make EE mandatory across all educational levels underscores the critical role of cross-sectoral networks in bridging the gap between environmental awareness and sustained behavioral change at scale. This finding suggests that effective scaling requires not only educational networks but also institutional partnerships that can translate awareness into systemic policy change.

10.3.5 Fidelity and Adaptation Assessment

Successful scaling requires evaluation systems that can monitor both fidelity to core principles and the appropriateness of contextual adaptations. Duong and Ngo (2025), in their mediation-moderation analysis of the relationships between ecotourism, green construction, EE, and sustainable behavior in Vietnam, provide a methodological model that illustrates how integrated evaluation approaches can simultaneously assess multiple program components and their interactions. Their finding that green building mediates the relationship between EE and sustainable development outcomes, while ecotourism moderates the relationship between green social behavior and sustainability goals and demonstrates the value of evaluation frameworks that capture complex interdependencies between program elements—a capacity essential for distinguishing between contextual adaptations that strengthen programs and those that may compromise their effectiveness.

10.4 RESOURCE MANAGEMENT AND FINANCIAL SUSTAINABILITY

10.4.1 Diversified Financing Models

The financial sustainability of EE programs requires developing diversified financing models that reduce dependence on single funding sources and create more resilient financing portfolios (Filho et al., 2020). These models should combine public, private, philanthropic, and self-generated resources within implementing organizations.

Carvalho Coutinho et al. (2024) investigated the public management and visitor perceptions of EE at Cocó State Park, Ceará, Brazil, highlighting the challenges associated with resource management in protected areas. The study reveals that while the park benefits from partnerships and interactive activities, it lacks brochures or booklets for distribution to volunteers and an effective system for counting visitors.

The results suggest that the Cocó State Park fulfills its role in EE but needs some adjustments to improve. The study identified the need to diversify resources to include educational materials, volunteer programs, and effective monitoring systems.

10.4.2 Strategies for Generating One's Own Income

Long-term sustainability requires organizations to develop the capacity to generate income to supplement external funding. These strategies may include the sale of educational services, consulting, educational products, and specialized training programs, among others.

Belcavello et al. (2024), in their research on massive open online courses (MOOCs) as a tool for promoting spring conservation and restoration applied to EE, illustrate innovative approaches to resource generation. Their development of a MOOC focused on EE, specifically spring conservation and preservation, demonstrates how digital technologies can amplify reach while optimizing costs.

This study highlighted the viability and positive impact of MOOCs in promoting inclusive and globally accessible EE. This model can generate revenue through certification, specialized services, and content licensing.

10.4.3 Efficient Management of Operational Resources

Programmatic effectiveness requires management systems that optimize the use of available resources while maintaining the quality of educational service. This management includes optimizing human resources, using infrastructure efficiently, and implementing procurement systems that maximize the value of the investment.

Efficient resource management in EE requires optimizing available resources while maintaining educational quality and community relevance. Kopnina (2020) argued that the most resource-efficient EE programs are those that leverage existing community structures, local expertise, and place-based learning opportunities rather than depending entirely on external resources and imported materials. Her critical evaluation found that programs designed around locally available resources—including natural environments, community knowledge holders, and existing educational infrastructure—consistently achieve better cost-effectiveness ratios and produce stronger connections between educational content and students' lived experiences. These findings suggest that efficient resource management begins at the design phase, with programs structured to maximize the educational value of locally accessible resources.

10.4.4 Strategic Partnerships for Sustainability

The development of strategic partnerships can provide access to complementary resources, technical expertise, and co-financing opportunities, thereby strengthening program financial sustainability. These partnerships require the alignment of objectives, clear distribution of responsibilities, and collaborative governance mechanisms.

Strategic partnerships between different knowledge systems can provide access to complementary resources and perspectives that strengthen EE programs. Brundiers et al. (2021) identified key competencies in sustainability education that emerge most effectively from partnerships between diverse institutional actors, including educational institutions, community organizations, indigenous knowledge holders, and scientific research centers. Their agreed-upon reference framework emphasizes that transdisciplinary partnerships—those that bridge different knowledge traditions and institutional cultures—are essential for developing EE programs that can cultivate globally eco-conscious citizens while respecting and integrating local knowledge systems. The authors concluded that the most effective partnerships are characterized by clear governance structures, mutual respect between knowledge traditions, and shared commitment to educational outcomes that serve both local community needs and broader sustainability goals.

10.4.5 Efficiency and Return on Investment Indicators

Effective resource management requires indicator systems that can measure the operational efficiency and social return on investment of EE programs (Veidemane, 2022). These systems must capture both direct costs and social and environmental benefits that can justify the continued investment.

Lee et al. (2023) present an analysis of effective EE for sustainable development in South Korea, particularly in the context of moving beyond the Plastic Age. Their findings indicate that the current EE curriculum is predominantly focused on the elementary level and fails to adequately convey to the public the specifics of "what" and "how" to live sustainably.

The study suggested that "the effectiveness of environmental education can be improved if: (1) guidance on how to live in harmony with the natural environment is incorporated for cooperation toward building a resilient and sustainable community with a clear purpose, (2) activities to build an environmentally friendly culture are reinforced, and (3) content is updated regularly."

10.4.6 Financial Accountability Mechanisms

Financial sustainability requires accountability systems that demonstrate responsible resource use and build trust among funders, beneficiaries, and other stakeholders. These systems must balance transparency with administrative efficiency, avoiding excessive bureaucratic burdens that can reduce the resources available for programmatic activities.

Financial accountability systems must capture not only conventional financial metrics but also the broader educational and environmental value generated by programs. Wamsler et al. (2021) argued that comprehensive accountability in sustainability education requires frameworks that integrate multiple dimensions of value, including cognitive learning outcomes, affective and relational transformations, behavioral changes, and longer-term contributions to environmental and community well-being. Their analysis proposed that accountability systems should adopt holistic and experiential approaches that recognize the interconnected nature of educational, social, and environmental outcomes and avoid reductive metrics that capture only easily quantifiable outputs while overlooking the deeper transformative processes that constitute the most significant contributions of EE programs.

10.5 CONCLUSION

10.5.1 Summary of Implementation Challenges

Analysis of effective implementation strategies reveals that the transition from theory to practice in EE presents multidimensional challenges requiring systemic, integrated approaches. Challenges include balancing fidelity to fundamental pedagogical principles with adaptation to specific contexts, coordinating multiple stakeholders with different objectives and capacities, and sustaining programs for sufficiently long periods to generate transformative impact.

The evidence presented demonstrates that successful implementation requires robust conceptual frameworks that can articulate specific theories of change while maintaining the flexibility for adaptation. Gilleran Stephens et al. (2025) illustrated how UDL can improve both inclusion and educational effectiveness. Similarly, Wu et al. (2025) found that evidence from the mechanisms underlying EE underscores the importance of explicit causal models in guiding design and implementation.

10.5.2 Critical Importance of Contextualization

A fundamental conclusion is that effective contextualization entails more than superficial adaptation of content or methodologies. Authentic contextualization requires a deep understanding of local socioecological systems, cultural structures, political dynamics, and economic conditions that influence the viability and effectiveness of various educational approaches.

Fernández-Llamazares et al. (2021), in their analysis of threats to indigenous and local knowledge systems, demonstrated how integrating traditional ecological knowledge into EE can enrich educational programs while strengthening local knowledge systems. Dodson and Miru (2021) provide compelling evidence from New Zealand that culturally appropriate methodologies can significantly advance indigenous self-determination while producing meaningful EE outcomes.

10.5.3 Critical Factors for Sustainable Scaling

Successful scaling of effective initiatives requires the systematic identification of essential elements to preserve during expansion and components that can be adapted to new contexts. Evidence suggests that the most effective scaling occurs when organizations can maintain fundamental principles while developing the capacity for systematic contextual adaptation.

Sobaih et al. (2024) present a study on mediating waste minimization behavior, offering a model that can be replicated in other higher education institutions.

10.5.4 Integrating Financial and Programmatic Sustainability

A key perspective is that financial sustainability and programmatic quality are not competing goals but complementary dimensions that require an integrated approach. The most sustainable programs are those that can demonstrate an effective impact while developing diversified funding models that reduce vulnerability to changes in the availability of specific resources.

Belcavello et al. (2024) demonstrated how digital technologies can amplify reach while optimizing costs in MOOCs for spring conservation. Ostrem and Hvenegaard (2025) illustrated how interagency collaboration can strengthen both programmatic and financial sustainability.

10.5.5 Need for Systemic Approaches

The analysis revealed that the most effective implementation occurs when programs are conceptualized as components of broader systems that include multiple actors, intervention levels, and time scales. This systemic perspective requires coordination frameworks that can align the efforts of different actors while respecting organizational autonomy and the diversity of approaches.

Duong and Ngo (2025) researched the integration of ecotourism, green building, EE, and sustainable behavior, providing evidence of the synergistic effects of integrated approaches. Kola-Olusanya et al. (2023) conducted research on EE for water pollution prevention, underscoring the need for cross-sector coordination to maximize impact.

10.5.6 Directions for Future Innovation

Based on the evidence analyzed, several priority areas for innovation in the implementation of EE have emerged. First, the development of digital technologies and platforms that can democratize access to high-quality EE while maintaining local and contextual relevance is necessary. Fraisl et al. (2022) demonstrated that citizen-science platforms are among the most successful models for leveraging digital technologies to simultaneously advance environmental research and education at global scales while maintaining local relevance.

Second, the construction of evaluation and research systems that can generate more robust evidence on the effectiveness of different implementation approaches across diverse contexts is required to provide evidence-based guidance for future initiatives.

Third, strengthening organizational capacities for adaptive management enables organizations to navigate the tension between program fidelity and contextual flexibility effectively. Olsson et al. (2022) provide a methodological model for this type of adaptive assessment through their longitudinal tracking of sustainability outcomes, demonstrating that continuous monitoring and iterative program adjustment are essential for maintaining educational effectiveness over extended implementation periods.

10.5.7 Imperative for Coordinated Action

This evidence constitutes a clear call for coordinated action among researchers, practitioners, policymakers, and funders to develop and implement more effective strategies (Rousell & Cutter-Mackenzie-Knowles, 2020). This coordination should focus on building shared knowledge of what works, for whom, and under what conditions.

The urgency of contemporary environmental challenges requires coordinated action to proceed quickly, but the complexity of effective implementation also demands care and rigor in developing approaches. Balancing urgency and quality is a fundamental challenge for leadership in EE.

Finally, the success of implementation efforts depends critically on sustained investment in human and institutional capacity building to sustain high-quality programs over decades to generate meaningful environmental transformations. This investment requires commitment from multiple sectors and recognition that effective EE is both a moral imperative and a practical necessity for sustainable development.

REFERENCES

Ardoin, N. M., Bowers, A. W., & Gaillard, E. (2020). Environmental education outcomes for conservation: A systematic review. *Biological Conservation*, *241*, 108224. https://doi.org/10.1016/j.biocon.2019.108224

Belcavello, D. A. B., Barbosa, M. P., Jesus, J. S. D., & Sgarbi, A. D. (2024, June 19–21). MOOC as an instrument for promoting the conservation and restoration of springs applied to environmental education. 2024 International Symposium on Computers in Education (SIIE), Coruña, Spain.

Brundiers, K., Barth, M., Cebrián, G., Cohen, M., Diaz, L., Doucette-Remington, S., Dripps, W., Habron, G., Harré, N., Jarchow, M., Losch, K., Michel, J., Mochizuki, Y., Rieckmann, M., Parnell, R., Walker, P., & Zint, M. (2021). Key competencies in sustainability in higher education—toward an agreed-upon reference framework. *Sustainability Science*, *16*(1), 13–29. https://doi.org/10.1007/s11625-020-00838-2

Carvalho Coutinho, D. L., Bezerra Ramos, V., Farias Pinheiro, L., Martins de Santiago, P. L., Almeida Silva, J., da Silva Lima Ribeiro, A. L., & Lucena, E. M. P. D. (2024). Gestão Pública e Percepção dos Visitantes Sobre a Educação Ambiental no Parque Estadual do Cocó, Ceará, Brasil. *Revista Brasileira de Geografia Física*, *17*(6), 4229–4245. https://doi.org/10.26848/rbgf.v17.6.p4229-4245

Chávez Santos, R., Carrasco Rituay, A. M., Rascón, J., Reina Marín, Y., Sánchez Bardales, E., Chávez Santos, Y. T., & Carranza Guevara, R. (2024). The transformative power of environmental education in the consumption of safety: The case of a popular housing estate. *Qubahan Academic Journal*, *4*(3), 440–453. https://doi.org/10.48161/qaj.v4n3a838

Chawla, L. (2020). Childhood nature connection and constructive hope: A review of research on connecting with nature and coping with environmental loss. *People and Nature*, *2*(3), 619–642. https://doi.org/10.1002/pan3.10128

Dodson, G., & Miru, M. (2021). Ngā Waihotanga Iho: Self-determination through Indigenous environmental education in New Zealand. *Australian Journal of Environmental Education*, *37*(3), 254–265. https://doi.org/10.1017/aee.2021.5

Duong, K. D., & Ngo, T. Q. (2025). Do ecotourism, green construction, environmental education, and sustainable behaviour lead to sustainable development? A mediation–moderation approach. *Environment, Development and Sustainability*, *27*(7), 16015–16032. https://doi.org/10.1007/s10668-024-04568-8

Fernández-Llamazares, Á., Lepofsky, D., Lertzman, K., Armstrong, C. G., Brondizio, E. S., Gavin, M. C., Lyver, P. O. B., Nicholas, G. P., Pascua, P., Reo, N. J., Reyes-García, V., Turner, N., Yletyinen, J., Anderson, E. N., Balée, W., Cariño, J., David-Chavez, D. M., Dunn, C. P., Garnett, S. C., Greening, S., Shain, J., Kuhnlein, H., Molnár, Z., Odonne, G., Retter, G. B., Ripple, W. J., Sáfián, L., Bahraman, A. S., Torrents-Ticó, M., & Vaughan, M. B. (2021). Scientists' warning to humanity on threats to indigenous and local knowledge systems. *Journal of Ethnobiology*, *41*(2), 144–169. https://doi.org/10.2993/0278-0771-41.2.144

Filho, W. L., Lange, A., Pretorius, R. W., Londero, L., Manolas, E., Alves, F., Azeiteiro, U., Rogers, J., Shiel, C., & Do Paco, A. (2020). *Universities as Living Labs for Sustainable Development—Supporting the Implementation of the Sustainable Development Goals*. Springer Cham. https://doi.org/10.1007/978-3-030-15604-6

Fraisl, D., Hager, G., Bedessem, B., Gold, M., Hsing, P.-Y., Danielsen, F., Hitchcock, C. B., Hulbert, J. M., Piera, J., Spiers, H., Thiel, M., & Haklay, M. (2022). Citizen science in environmental and ecological sciences. *Nature Reviews Methods Primers*, *2*(1), 64. https://doi.org/10.1038/s43586-022-00144-4

Gal, A. (2024). Strengths, weaknesses, opportunities and threats: A SWOT analysis of a long-term outdoor environmental education program in Israel. *Journal of Outdoor and Environmental Education*, *27*(2), 321–339. https://doi.org/10.1007/s42322-023-00125-5

Gilleran Stephens, C., Antwi, S. H., & Linnane, S. (2025). Universal Design for Learning (UDL): A framework for re-design of an Environmental Education (EE) outreach program for a more inclusive and impactful Science Festival event. *Discover Education*, *4*(1), 217. https://doi.org/10.1007/s44217-025-00660-x

Kola-Olusanya, A., Oyeyemi, E., Adewale, P. S., & Omobuwa, O. (2023). Role of environmental education in water pollution prevention and conservation in Nigeria. *Water Supply*, *24*(2), 361–370. https://doi.org/10.2166/ws.2023.337

Kopnina, H. (2020). Education for the future? Critical evaluation of education for sustainable development goals. *The Journal of Environmental Education*, *51*(4), 280–291. https://doi.org/10.1080/00958964.2019.1710444

Lee, M., Yang, J.-E., Kim, H., Moon, J., Ryu, H.-S., & Lee, J.-Y. (2023). Effective environmental education for sustainable development beyond the Plastic Age in South Korea. *Environmental Education Research*, *29*(9), 1328–1345. https://doi.org/10.1080/13504622.2023.2190530

Lozano, R., Barreiro-Gen, M., Lozano, F. J., & Sammalisto, K. (2019). Teaching sustainability in European higher education institutions: Assessing the connections between competences and pedagogical approaches. *Sustainability*, *11*(6), 1602. https://doi.org/10.3390/su11061602

Menton, M., Larrea, C., Latorre, S., Martinez-Alier, J., Peck, M., Temper, L., & Walter, M. (2020). Environmental justice and the SDGs: From synergies to gaps and contradictions. *Sustainability Science*, *15*(6), 1621–1636. https://doi.org/10.1007/s11625-020-00789-8

Olsson, D., Gericke, N., & Boeve-de Pauw, J. (2022). The effectiveness of education for sustainable development revisited: A longitudinal study on secondary students' action competence for sustainability. *Environmental Education Research*, *28*(3), 405–429. https://doi.org/10.1080/13504622.2022.2033170

Ostrem, J., & Hvenegaard, G. (2025). Interagency collaboration for environmental education: Insights from the Beaver Hills Biosphere, Canada. *Journal of Environmental Planning and Management, 68*(1), 227–244. https://doi.org/10.1080/09640568.2023.2240954

Redman, E., Wiek, A., & Redman, A. (2018). Continuing professional development in sustainability education for K-12 teachers: Principles, programme, applications, outlook. *Journal of Education for Sustainable Development, 12*(1), 59–80. https://doi.org/10.1177/2455133318777182

Rousell, D., & Cutter-Mackenzie-Knowles, A. (2020). A systematic review of climate change education: Giving children and young people a 'voice' and a 'hand' in redressing climate change. *Children's Geographies, 18*(2), 191–208. https://doi.org/10.1080/14733285.2019.1614532

Sobaih, A., Gharbi, H., Zaiem, I., & Aliane, N. (2024). ROSE (Recycling Organization through Sustainability Education): Examining the mediating effects of waste minimization behaviour in the relationship between environmental education and environmental performance at the Saudi Arabian Universities. *Journal of Infrastructure, Policy and Development, 8*(6), 6462. https://doi.org/10.24294/jipd.v8i6.6462

Veidemane, A. (2022). Education for sustainable development in higher education rankings: Challenges and opportunities for developing internationally comparable indicators. *Sustainability, 14*(9), 5102. https://doi.org/10.3390/su14095102

Wamsler, C., Osberg, G., Osika, W., Herndersson, H., & Mundaca, L. (2021). Linking internal and external transformation for sustainability and climate action: Towards a new research and policy agenda. *Global Environmental Change, 71*, 102373. https://doi.org/10.1016/j.gloenvcha.2021.102373

Whitburn, J., Linklater, W., & Abrahamse, W. (2020). Meta-analysis of human connection to nature and proenvironmental behavior. *Conservation Biology, 34*(1), 180–193. https://doi.org/10.1111/cobi.13381

Wu, X., Jia, W., & Wu, T. (2025). Mechanism by which environmental education influences pro-environmental behavior in Wuyishan National Park, China. *Sustainability, 17*(1), 0043. https://doi.org/10.3390/su17010043

11 Success Stories and Global Best Practices

11.1 NOTABLE PROGRAMS IN LATIN AMERICA

11.1.1 Green School Certification Initiatives in Chile

The green school certification program in Chile is an outstanding example of how public policies encourage the systematic integration of environmental education into formal educational institutions. Salazar et al. (2024) evaluated the effects of this program on children's pro-environmental behavior using "a multivalued treatment effects model to assess the effects of a green school certification program on children's pro-environmental behavior."

The results showed "a positive effect of higher school certification levels on children's behavior toward plastic waste." This effect appears to be stronger in practices in which children have greater decision-making authority, such as "packing a lunch box." Significantly, the study observed a "reverse effect when moving from the basic to the intermediate level of certification, which is in line with the potentially nonlinear effects of environmental education on pro-environmental behavior."

The Chilean program demonstrates the importance of designing differentiated incentive systems that recognize progressive achievements in school sustainability. The experience suggests that "to improve program design, it is important to redefine incentives in the certification system to differentiate better the benefits of achieving each level of environmental certification."

11.1.2 Primate Conservation Programs in Mexico

The primate conservation education program implemented in Los Tuxtlas, Mexico, illustrates how participatory and arts-based approaches can bring about significant transformations in local communities' perceptions and knowledge of endangered species. Research by Hernández-Cortés et al. (2024) documented a program targeting secondary school students that used "participatory action research and arts-based education."

The results indicated "a decrease in students' fear of monkeys, the recovery of myths and legends about this species, as well as a complex knowledge about monkeys and their habitat." The inclusion of local knowledge "enriched the learning experience and reinforced concepts of primate conservation." Participatory action research and arts-based approaches "enabled students to internalize and apply the knowledge acquired and fostered empathy."

The program also developed teamwork skills, critical thinking, and creativity. Quantitative indicators showed changes in emotions and knowledge after the program, while qualitative data revealed "a deep articulation of the content learned."

DOI: 10.1201/9781003779667-15

The program identified migration and mining activities as sociopolitical factors that impact local livelihoods and wildlife conservation in the region.

11.1.3 Water Management Experiences in Brazil

The Portugal–Brazil collaboration project on water reuse as an environmental education tool demonstrates how international partnerships can amplify the impact of environmental education initiatives. Research by Santos et al. (2024) presented "the Water Reuse Institute (IrdA) as a high-quality environmental education tool" that measured "the reach of the IrdA's direct and indirect educational actions."

The results showed that "direct actions proved effective in disseminating knowledge, highlighting a preference for the topic of 'regulation' and adherence to content produced in different video formats." Indirect actions showed that "the scientific approach is less accessible because of its complex language. However, when presented in more accessible data, scientific content gains trust and can be highly relevant to the transformation of society."

The findings corroborate "the current trend toward greater educational outreach in free and digital formats to build a society truly committed to water reuse and more efficient water management." This experience illustrates how digital platforms can democratize access to technical and scientific knowledge while maintaining academic quality.

11.1.4 Rural Environmental Education Initiatives in Colombia

Transformative environmental education programs implemented in rural Colombia have demonstrated the importance of connecting educational approaches with specific territorial realities. Ardoin et al. (2020), in their systematic review of environmental education outcomes, found that programs incorporating place-based and community-centered approaches consistently produced stronger conservation and behavioral outcomes, particularly in rural contexts where formal education has traditionally been disconnected from local environmental realities. These programs have sought to overcome the traditional disconnect between formal education and rural realities by integrating local knowledge and educational practices that encourage citizen participation in social transformation.

Colombian programs have emphasized human connections with territories and interdisciplinary management as central elements in their educational approaches. The integration of cultural ecology and rural educational processes has enabled the development of contextualized educational proposals that connect with specific environments, fostering meaningful learning by enriching environmental education with local elements.

11.1.5 Network of Environmental Education Centers in Costa Rica

Costa Rica has developed a national network of environmental education centers that combine biodiversity conservation with innovative educational programs. Chawla (2020), in her comprehensive review of research on nature connection, found that programs providing sustained direct contact with natural environments—such as those offered by dedicated environmental education centers—consistently foster stronger emotional connections to nature and more enduring conservation commitments among participants. These centers have implemented approaches that integrate sustainable tourism with environmental education, creating self-financing models that demonstrate long-term sustainability.

The programs have integrated the local ecological knowledge of fishermen with the scientific principles of marine conservation, creating educational syntheses that have strengthened both sustainable fishing practices and the scientific understanding of marine ecosystems.

11.2 INTERNATIONAL REFERENCE EXPERIENCES

11.2.1 Environmental Education Programs in European Zoos

Feucht et al. (2025) analyzed educational programs at nine zoos using pre- and post-questionnaires. The results showed that the adjusted instruments worked well with some adaptations and that interest in species conservation could be measured as a general six-item construct. The study also found that environmental concern was three-dimensional, as intended, and could distinguish between selfish, biospheric, and altruistic concerns in zoos.

European zoos have developed programs that use direct encounters with wildlife to foster emotional connections that can be translated into conservation behaviors. The scales developed can be used for assessment purposes and are even applicable beyond zoos, suggesting that the methodological frameworks developed in these contexts have the potential to be transferred to other environmental education institutions.

11.2.2 Eco-Literacy Initiatives in Kenya

Environmental education programs implemented on the Kenyan coast illustrate how tourism-promoted initiatives can generate significant educational impacts in local communities in Kenya. Berman (2021) explored select German initiatives in the education sector that have emerged in the context of the area's tourism industry.

The findings revealed a lesser-known dimension of tourism, namely, the successful pursuit of education for sustainable development (ESD) in schools that thrive in the context of tourism through an integrated approach to teaching eco-literacy. The schools analyzed serve a significant portion of the population of the wider Diani-Ukunda area, including the interior that extends into Kwale County.

The study found that schools fulfill their role in environmental education but require some adjustments to improve it. The Kenyan experience demonstrates how international funding networks based on tourism contacts can sustain long-term educational initiatives that integrate conservation objectives and community development.

11.2.3 Educational Conservation Programs in Madagascar

Madagascar has developed innovative programs that integrate environmental education and biodiversity conservation. Brias-Guinart et al. (2023) compared practitioners' perspectives on the purposes of environmental education in Finland and Madagascar, finding significant differences in their approaches and priorities.

Practitioners in Madagascar place greater emphasis on the direct links between environmental education and community survival, reflecting the socioeconomic realities in which conservation and human well-being are closely connected. Madagascan programs have developed approaches that integrate the conservation of endemic species with community capacity building for sustainable natural resource management.

Madagascar has been particularly innovative in developing educational materials that use local narratives and traditional cultural elements to communicate conservation concepts. The programs have demonstrated that culturally rooted approaches can be more effective than imported methodologies in generating lasting changes in resource-use practices.

11.2.4 Israeli Model of Outdoor Environmental Education

Israel has developed long-term outdoor environmental education programs that demonstrate exceptional institutional sustainability. Research by Gal (2024) analyzed a program implemented continuously since 1996, finding that "the strengths and opportunities seem to outweigh the threats and weaknesses of such a program, enabling the long-term success of the program."

The strengths, weaknesses, opportunities, and threats (SWOT) analysis revealed that long-term outdoor environmental education programs that enable the achievement of environmental education goals over time can have significant positive impacts, both environmentally and on the communities involved, and are sustainable when appropriate attention is paid to relationships with experts and funding agencies.

The Israeli model has been successful in maintaining programmatic consistency over decades while adapting to changes in the political, social, and environmental contexts. Experience demonstrates the importance of building strong institutional relationships and diversified funding systems for sustainability.

11.2.5 Approaches to Environmental Education in Finland

Finland has developed unique approaches to ecological education that emphasize connections to the Nordic environment and integrate sustainability principles into state educational structures. As Brias-Guinart et al. (2023) noted in their comparative study of environmental education practitioners' perspectives in Finland and Madagascar, Finnish programs are characterized by their systematic integration of sustainability principles into state educational structures and their emphasis on holistic approaches that connect environmental learning with broader social welfare objectives. Finnish programs have been praised worldwide for their ability to combine ecological education with broader social welfare and civic skills development.

The Finnish experience demonstrates the value of comprehensive approaches that link environmental education with health education, civic education, and life skills development. Olsson et al. (2022) provide longitudinal evidence from Nordic educational contexts demonstrating that sustained, curriculum-integrated sustainability education produces measurable improvements in students' sustainability consciousness over time, supporting the Finnish model of embedding environmental education within the core educational framework rather than treating it as supplementary content. The plans have used many natural sites as study locations, leveraging the abundance of nearby natural areas to create immersive learning experiences for students.

11.3 SUCCESSFUL PEDAGOGICAL AND TECHNOLOGICAL INNOVATIONS

11.3.1 Augmented Reality in Environmental Education

The integration of augmented reality technologies into environmental education has demonstrated significant potential to create immersive learning experiences that enhance the impact of traditional projects. Simon et al. (2025) conducted a systematic review of environmental research to identify pathways and options in this new field.

The use of augmented reality has made it easier for students to visualize complex environmental processes, discover distant natural sites, and experiment with replicas of future scenarios. These technologies have been very effective in making abstract ideas such as climate change, biogeochemical cycles, and dynamics of natural systems, which have always been difficult to convey through conventional means, more tangible.

Programs that incorporate augmented reality have shown significant increases in student interest, knowledge retention, and the transfer of learning to real-world settings. However, successful implementation requires careful attention to issues such as technological accessibility, teacher training, and pedagogical integration.

11.3.2 Chatbots and Games with Environmentally Conscious Design

The advancement of chatbots specifically designed for environmental education is a promising development because it integrates artificial intelligence with pedagogical approaches. Nguyen et al. (2025) explained how design is sensitive to the values of chatbots in environmental education, supporting identity, connectivity, well-being, and sustainability.

The chatbots that have integrated ethical and environmental values into their interaction patterns extend learning beyond information transmission to encompass reflection on personal value and how one is committed to caring for the planet. These systems have demonstrated the ability to adapt their speech to each person's values, making conversations more meaningful and useful for the user.

New chatbot features that teach about the environment include facilitating discussions on ethical environmental issues, providing personalized guidance on pro-environmental behavior, and connecting users with locally relevant opportunities for environmental action. These systems represent an effective integration of environmental education and artificial intelligence.

Serious games for environmental sustainability have become valuable tools for environmental education, combining the entertainment value of video games with the rigor of learning essential concepts in the field. Ahmadov et al. (2025) conducted a two-phase systematic literature review on the use of serious games for sustainable environmental education. These games have proven effective in teaching decision-making in complex environmental situations, allowing players to explore different approaches to caring for the planet and observe the consequences of their decisions in simulated environments without risk.

Notable improvements in educational games have combined engaging stories with playful systems that simulate real environmental movements, resulting in experiences that are both educational and engaging. Ideal cases have demonstrated the ability to transfer learning from game situations to real-world activities.

11.3.3 Interactive Computer-Based Training

Computer-based interactive training programs have expanded the possibilities for nature-based environmental education, particularly in contexts where access to natural spaces is restricted. Dong (2025) examined the impact of these technologies on conservation behaviors, social behaviors, and lower- and higher-order thinking skills.

The results showed that the program significantly improved higher-order thinking skills, conservation behaviors, and social behaviors in both the technology-using and control groups. Interestingly, no significant differences were found between the two approaches to project participation: With and without additional video viewing/online discussions.

This evidence suggests that while interactive technologies can enrich educational experiences, the program's fundamental pedagogical design may be more important than specific technological sophistication. The most successful programs have used technology to amplify strong pedagogical approaches rather than as a substitute for thoughtful educational design.

11.3.4 Digital Escape Rooms for Environmental Education

Digital escape rooms represent a methodological innovation that combines elements of gamification and collaborative learning to create immersive and memorable educational experiences. Ouariachi et al. (2020), in their systematic review of gamification approaches for pro-environmental education, identified key design principles that underpin effective environmental games, including narrative framing, collaborative mechanics, and feedback systems that connect participant decisions to environmental consequences. Their analysis provides the theoretical foundation for understanding why digital escape rooms—which inherently incorporate these design principles—have proven effective as environmental education tools.

Environmental escape rooms have demonstrated their ability to engage participants in solving complex problems that require the integration of diverse environmental knowledge. These experiences have been particularly effective in developing teamwork skills, critical thinking, and the practical application of environmental knowledge in simulated contexts that require decision-making under pressure. Digital escape rooms have made it possible to create elaborate environmental scenarios that are impossible to recreate in physical settings.

The most successful implementations have carefully combined specific educational objectives with escape room methods, ensuring that solving the escape rooms requires a true grasp of environmental concepts rather than just the skills of ordinary players.

11.3.5 Mobile Learning Platforms

The development of mobile applications dedicated to environmental education has democratized access to high-quality educational experiences, enabling ubiquitous learning across multiple contexts and times. Petersen et al. (2020) demonstrated that technology-mediated environmental learning experiences, including those delivered through mobile platforms, can significantly enhance knowledge acquisition and attitudinal change when designed according to evidence-based multimedia learning principles. The most innovative applications have integrated geolocation capabilities to provide real-time, location-specific environmental information.

Mobile platforms have proven particularly valuable in citizen-science programs because they enable participants to collect environmental data while learning about the ecological processes in their surroundings. Fraisl et al. (2022), in their comprehensive review of citizen science in environmental and ecological sciences published in Nature Reviews Methods Primers, confirmed that mobile-enabled citizen-science platforms simultaneously advance scientific research and provide powerful learning experiences that connect participants to authentic ecological inquiry at unprecedented scales. In addition, these applications directly connect individual learning to tangible contributions to scientific research.

11.4 LESSONS LEARNED AND CRITICAL SUCCESS FACTORS

11.4.1 Importance of Cultural and Territorial Contextualization

A key lesson from successful cases is the importance of adapting environmental education plans to the specific characteristics of the areas, cultures, and groups in which they are implemented. The best plans have demonstrated the ability to combine general environmental education principles with precise adjustments that reflect local contexts.

Atuguba and Yakpir (2024), in their work "On Ghanaian University Teachers' Experiences and Perceptions of Traditional Environmental Beliefs and Practices: Implications for Environmental Education," underscore the significance of this contextualization. Their study found that respectfully integrating traditional knowledge systems can significantly enrich environmental education programs while strengthening local cultural identities.

Programs that have successfully integrated traditional and scientific knowledge have developed specific methodologies to facilitate genuine dialogue between different forms of knowledge, avoiding both the rejection of traditional knowledge and the uncritical romanticization of ancestral practices.

11.4.2 Sustainable Institutional Capacity Building

The success stories analyzed demonstrate that the long-term sustainability of environmental education programs critically depends on the development of robust institutional capacities that can maintain program quality over extended periods, adapt to changes in operating contexts, and generate the resources necessary for continuity in the program.

Brundiers et al. (2021) developed an agreed-upon reference framework for key competencies in sustainability education in higher education, providing critical insights into the institutional factors that facilitate long-term program sustainability. Their framework, developed through extensive international consultation with sustainability education experts, emphasizes the importance of clear conceptual foundations, well-defined competency systems aligned with international policy frameworks, and institutional structures that support the continuous development of educator and learner capacities. The authors identified systems thinking, anticipatory competence, normative competence, strategic competence, and interpersonal competence as essential for building sustainable environmental education programs. They concluded that sustainable programs require competency frameworks that are simultaneously rigorous enough to ensure educational quality and flexible enough to accommodate diverse institutional and cultural contexts.

The most sustainable programs have invested significantly in human capacity development, creating robust organizational systems, and constructing support networks that can provide technical and financial resources over extended periods.

11.4.3 Integration of Multiple Stakeholders

A consistent feature of successful programs is their ability to effectively articulate the perspectives of multiple stakeholders with different capacities, resources, and perspectives on shared environmental education goals. This integration requires the development of collaborative governance frameworks that align diverse incentives while respecting organizational autonomy.

Reilly et al. (2022), in the paper entitled "Agricultural and Environmental Education: A Call for Meaningful Collaboration in a U.S context," analyzed the context and illustrated the potential benefits and challenges of cross-sector integration. Their analysis suggests that effective collaboration requires "breaking down existing disciplinary walls between AE [agricultural education] and EE [environmental education] to reposition both as critical public goods effectively."

Programs that have achieved successful integration have developed specific mechanisms to facilitate dialogue, conflict resolution, equitable benefit distribution, and participatory decision-making that accommodate diverse perspectives.

11.4.4 Balancing Scientific Rigor and Accessibility

Successful cases have demonstrated the importance of balancing scientific rigor with communicative accessibility, ensuring that educational programs maintain technical accuracy while remaining understandable and relevant to diverse audiences. This balance requires the development of specific competencies to translate complex scientific knowledge into appropriate educational formats.

Schlickmann and Bortoluzzi (2023), in the paper "Evaluating Environmental Education Performance in a Higher Education Institution," provide a framework for assessing this balance. Their analysis emphasizes that successful institutions develop capabilities to "communicate research results in an accessible manner without compromising scientific rigor."

The most effective programs have developed multiple communication strategies to reach different audiences with varying levels of prior knowledge using narratives, visualizations, hands-on experiences, and other approaches to make complex scientific information accessible.

11.4.5 Impact Measurement and Evaluation

A central notion is that it is advisable to create robust measurement and assessment systems to capture both the rapid effects and profound changes that arise over time. Typically, programs with the greatest impact invest heavily in comprehensive assessment systems that integrate quantitative data (to measure scale and trajectories) and qualitative data (to understand experiences, steps, and meanings).

A clear case of orderly evaluation is the analysis by Wang et al. (2022) of the effect of studying the local environment on tourism practices. Their idea brought together questions, methods for extracting metaphors, and structural account models, enabling the examination of complex effects from multiple perspectives and in greater depth than previously possible.

In addition, the most rigorous assessment systems often include participatory elements: They involve program participants and beneficiaries in defining what "success" means and in explaining the effects. This not only improves the assessment but also strengthens the community's capacity for self-assessment and identifies avenues for improvement.

11.4.6 Diversified Financing and Economic Sustainability

Successful cases have demonstrated the importance of developing diversified financing models to reduce dependence on single funding sources and build financial resilience. Filho et al. (2020), in their analysis of universities functioning as living laboratories for sustainable development, demonstrated that institutions achieving long-term financial sustainability in their environmental education programs consistently develop diversified funding portfolios that combine public grants, private partnerships, philanthropic support, and self-generated income from educational services and consulting activities. The most sustainable programs combine public, private, and philanthropic funding with self-generated income.

11.5 CONCLUSION

11.5.1 Diversity of Successful Approaches

An analysis of global success stories and best practices reveals a remarkable diversity of approaches that have proven effective across contexts, confirming that there is no single model for successful environmental education but rather multiple paths toward the shared goals of social and environmental transformation. This diversity reflects the richness of global pedagogical traditions and the need to adapt educational approaches to specific contextual realities.

The cases analyzed show that success in environmental education can manifest through school certification programs, as in Chile; community conservation initiatives, as in Mexico; collaborative digital platforms, as in Brazil; long-term outdoor education programs, as in Israel; or technological innovations such as escape digital rooms and augmented reality applications. This diversity suggests that the field of environmental education has developed a rich toolbox adaptable to diverse contexts and needs.

11.5.2 Cross-Cutting Success Factors

Despite the diversity of specific approaches, cross-cutting factors emerge that characterize successful programs across contexts. These factors include authentic commitment to community participation, respectful integration of multiple forms of knowledge, development of sustainable institutional capacity, and robust systems for evaluation and continuous improvement.

Menton et al. (2020) emphasize that successful environmental education programs must systematically address issues of equity and environmental justice, ensuring that evaluation processes consider how factors such as socioeconomic status, cultural background, race, and geographic location influence both program implementation and the validity of evaluation findings. Their analysis of environmental justice dimensions across diverse international contexts demonstrates that genuine success in environmental education requires attention not only to educational outcomes but also to the equitable distribution of both environmental benefits and educational opportunities. The authors concluded that evaluation frameworks that fail to account for structural inequities risk producing assessments that validate programs serving privileged populations while overlooking the needs of marginalized communities most affected by environmental degradation.

Cross-cutting factors also include the ability to balance scientific rigor with communicative accessibility, the development of strategic partnerships that expand available resources and capacities, and the construction of diversified funding systems that can sustain programs over extended periods necessary to generate profound transformations.

11.5.3 Importance of Continuous Innovation

Successful cases highlight the importance of maintaining a focus on continuous innovation, enabling programs to adapt to changes in technological, social, and environmental contexts. The most impactful innovations have emerged when organizations combine a deep understanding of fundamental educational principles with openness to experimentation with new methodologies, technologies, and organizational approaches.

Mullenbach and Green (2018) illustrate how programs can innovate by targeting specific audiences that have traditionally been underrepresented in environmental education. Their study found that innovative approaches can effectively reach diverse demographic groups when appropriately tailored to their specific characteristics and contexts. This finding is consistent with the broader evidence presented by Ardoin et al. (2020), whose systematic review confirmed that environmental education programs demonstrating the greatest innovation in reaching underserved populations consistently produced stronger and more equitable conservation outcomes across diverse demographic groups.

Successful innovation in environmental education requires a careful balance between experimentation and rigor, ensuring that new approaches remain faithful to fundamental educational objectives while exploring creative possibilities for greater effectiveness and impact.

11.5.4 Transferability and Contextual Adaptation

An important lesson is that the transferability of good practices across contexts requires careful adaptation processes that identify essential elements to preserve and components to modify based on specific contextual characteristics of the receiving context. Attempts at direct transfer without adaptation often result in less effective implementations or complete failures.

Poje et al. (2024) illustrate the importance of contextual adaptation by finding significant differences between specialized and regular ecological programs. Their study suggests that "it is necessary to include environmental education to a greater extent as a foundation in regular compulsory program curricula if we are to achieve some of the key objectives of sustainable development."

The most successful transfer processes have included phases of contextual diagnosis, collaborative adaptation involving local stakeholders, careful piloting with continuous evaluation, and iterative refinement based on emerging knowledge.

11.5.5 Future Directions for Field Development

An analysis of successful cases suggests several promising directions for the future development of environmental education. First, greater integration between environmental education and other educational areas, such as health, civic, and social justice education, can create synergies that amplify the impact while optimizing resources. Brundiers et al. (2021) identified this integration as essential in their competency framework, noting that sustainability challenges inherently require transdisciplinary educational approaches that connect environmental, social, and ethical dimensions.

Second, developing sophisticated evaluation frameworks that capture systemic impacts and long-term transformations is important for demonstrating value and justifying investments in environmental education. Olsson et al. (2022) provide a methodological model for this type of longitudinal assessment, demonstrating that tracking sustainability consciousness over extended periods reveals developmental patterns that short-term evaluations consistently miss. Third, exploring emerging technologies such as artificial intelligence, virtual reality, and adaptive learning platforms can open new possibilities for personalizing and scaling high-quality educational experiences.

11.5.6 Imperative for Coordinated Global Action

The success stories analyzed in this chapter underscore the imperative for more coordinated global action in environmental education. Such coordination would facilitate the systematic exchange of evidence-based practices, the transfer of methodological innovations across contexts, and collaborative approaches to shared environmental challenges. Achieving this coordination requires balancing respect for local contextual specificities with the strategic leverage of international networks to amplify collective impact.

The urgency of environmental challenges demands that the international environmental education community accelerate its cycles of learning and cooperation without compromising quality or local relevance. The cases analyzed demonstrate that this acceleration is achievable when institutions commit to systematic documentation, rigorous evaluation, and transparent sharing of both successes and failures.

Ultimately, these cases demonstrate that effective environmental education transcends technical implementation to embody an ethical commitment to intergenerational equity. The most impactful programs are those that have translated this commitment into sustained institutional practice, transforming the relationships between human communities and natural systems through education that is simultaneously rigorous, equitable, and contextually meaningful.

REFERENCES

Ahmadov, T., Karimov, A., Durst, S., Saarela, M., Gerstlberger, W., Wahl, M. F., & Karkkainen, T. (2025). A two-phase systematic literature review on the use of serious games for sustainable environmental education. *Interactive Learning Environments*, *33*(3), 1945–1966. https://doi.org/10.1080/10494820.2024.2414429

Ardoin, N. M., Bowers, A. W., & Gaillard, E. (2020). Environmental education outcomes for conservation: A systematic review. *Biological Conservation*, *241*, 108224. https://doi.org/10.1016/j.biocon.2019.108224

Atuguba, J. A., & Yakpir, G. M. (2024). Ghanaian college teachers' experiences and perceptions of traditional environmental beliefs and practices: Implications for environmental education. *The Journal of Environmental Education*, *55*(3), 238–248. https://doi.org/10.1080/00958964.2023.2286290

Berman, N. (2021). Environmental education catalyzed by tourism: Ecoliteracy initiatives on the coast of Kenya. *Sustainability*, *13*(15), 8501. https://doi.org/10.3390/su13158501

Brias-Guinart, A., Aivelo, T., Högmander, M., Heriniaina, R., & Cabeza, M. (2023). A better place for whom? Practitioners' perspectives on the purpose of environmental education in Finland and Madagascar. *The Journal of Environmental Education*, *54*(3), 163–180. https://doi.org/10.1080/00958964.2023.2178371

Brundiers, K., Barth, M., Cebrián, G., Cohen, M., Diaz, L., Doucette-Remington, S., Dripps, W., Habron, G., Harré, N., Jarchow, M., Losch, K., Michel, J., Mochizuki, Y., Rieckmann, M., Parnell, R., Walker, P., & Zint, M. (2021). Key competencies in sustainability in higher education—toward an agreed-upon reference framework. *Sustainability Science*, *16*(1), 13–29. https://doi.org/10.1007/s11625-020-00838-2

Chawla, L. (2020). Childhood nature connection and constructive hope: A review of research on connecting with nature and coping with environmental loss. *People and Nature*, *2*(3), 619–642. https://doi.org/10.1002/pan3.10128

Dong, Q. (2025). Computer-based interactive training in nature-based environmental education: Impact on conservation behaviours, social behaviours, and lower and higher order thinking skills. *Thinking Skills and Creativity*, *57*, 101825. https://doi.org/10.1016/j.tsc.2025.101825

Feucht, V., Dierkes, P. W., Haeser-Kalthoff, A., Kästner, A., Mager, C., Reichhardt, S., Schröder, L., Spengler, T., Steinert, I., Ugolini, P., Waked, F., & Kleespies, M. W. (2025). Environmental education in zoos: Analysis of different scales to measure the impact of educational programs. *Environmental Education Research*, *31*(4), 776–793. https://doi.org/10.1080/13504622.2024.2386631

Filho, W. L., Lange, A., Pretorius, R. W., Londero, L., Manolas, E., Alves, F., Azeiteiro, U., Rogers, J., Shiel, C., & Do Paco, A. (2020). *Universities as Living Labs for Sustainable Development—Supporting the Implementation of the Sustainable Development Goals*. Springer Cham. https://doi.org/10.1007/978-3-030-15604-6

Fraisl, D., Hager, G., Bedessem, B., Gold, M., Hsing, P.-Y., Danielsen, F., Hitchcock, C. B., Hulbert, J. M., Piera, J., Spiers, H., Thiel, M., & Haklay, M. (2022). Citizen science in environmental and ecological sciences. *Nature Reviews Methods Primers*, *2*(1), 64. https://doi.org/10.1038/s43586-022-00144-4

Gal, A. (2024). Strengths, weaknesses, opportunities and threats: A SWOT analysis of a long-term outdoor environmental education program in Israel. *Journal of Outdoor and Environmental Education*, *27*(2), 321–339. https://doi.org/10.1007/s42322-023-00125-5

Hernández-Cortés, A., Silva-Rivera, E., Rangel-Negrín, A., & Franquesa-Soler, M. (2024). Teenagers' perceptions of Mexican primates: A participatory environmental education program for the conservation of mantled howler monkeys (Alouatta palliata) in Los Tuxtlas, Mexico. *Folia Primatologica*, *95*(4–6), 325–336. https://doi.org/10.1163/14219980-bja10022

Menton, M., Larrea, C., Latorre, S., Martinez-Alier, J., Peck, M., Temper, L., & Walter, M. (2020). Environmental justice and the SDGs: From synergies to gaps and contradictions. *Sustainability Science*, *15*(6), 1621–1636. https://doi.org/10.1007/s11625-020-00789-8

Mullenbach, L. E., & Green, G. T. (2018). Can environmental education increase student-athletes' environmental behaviors? *Environmental Education Research*, *24*(3), 427–444. https://doi.org/10.1080/13504622.2016.1241218

Nguyen, H., Nguyen, V., Ludovise, S., & Santagata, R. (2025). Value-sensitive design of chatbots in environmental education: Supporting identity, connectedness, well-being and sustainability. *British Journal of Educational Technology*. https://doi.org/10.1111/bjet.13568

Olsson, D., Gericke, N., & Boeve-de Pauw, J. (2022). The effectiveness of education for sustainable development revisited: A longitudinal study on secondary students' action competence for sustainability. *Environmental Education Research*, *28*(3), 405–429. https://doi.org/10.1080/13504622.2022.2033170

Ouariachi, T., Li, C.-Y., & Elving, W. J. L. (2020). Gamification approaches for education and engagement on pro-environmental behaviors: Searching for best practices. *Sustainability*, *12*(11), 4565. https://doi.org/10.3390/su12114565

Petersen, G. B., Klingenberg, S., Mayer, R. E., & Makransky, G. (2020). The virtual field trip: Investigating how to optimize immersive virtual learning in climate change education. *British Journal of Educational Technology*, *51*(6), 2099–2115. https://doi.org/10.1111/bjet.12991

Poje, M., Marinić, I., Stanisavljević, A., & Rechner Dika, I. (2024). Environmental education on sustainable principles in kindergartens: A foundation or an option? *Sustainability*, *16*(7), 2707. https://doi.org/10.3390/su16072707

Reilly, C., Stevenson, K., Warner, W., Park, T., Knollenberg, W., Lawson, D., Brune, S., & Barbieri, C. (2022). Agricultural and environmental education: A call for meaningful collaboration in a U.S. context. *Environmental Education Research*, *28*(9), 1410–1422. https://doi.org/10.1080/13504622.2022.2040431

Salazar, C., Jaime, M., Leiva, M., & González, N. (2024). Environmental education and children's pro-environmental behavior on plastic waste: Evidence from the green school certification program in Chile. *International Journal of Educational Development*, *109*, 103106. https://doi.org/10.1016/j.ijedudev.2024.103106

Santos, A. S. P., Vieira, J. M. P., Lima, M. A. D. M., Soares, S. R. A., Pereira, L. V., Sampaio, V. M. P., Lima, E. L. D. S., Pavan, F. D. S., Lima, M. M. C. L., & Araujo, B. M. (2024). A technical–scientific content management system on water reuse as an environmental education tool: The experience of a Portugal/Brazil partnership. *Water Supply*, *24*(10), 3603–3616. https://doi.org/10.2166/ws.2024.214

Schlickmann, A., & Bortoluzzi, S. C. (2023). Environmental education performance evaluation in a higher education institution. *Systemic Practice and Action Research*, *36*(6), 935–965. https://doi.org/10.1007/s11213-023-09636-0

Simon, P. D., Zhong, Y., Cruz, I. C. D., & Fryer, L. K. (2025). Scoping review of research on augmented reality in environmental education. *Journal of Science Education and Technology*, *34*(4), 919–935. https://doi.org/10.1007/s10956-025-10218-z

Wang, J., Dai, J., Dewancker, B. J., Gao, W., Liu, Z., & Zhou, Y. (2022). Impact of situational environmental education on tourist behavior: A case study of water culture ecological park in China. *International Journal of Environmental Research and Public Health*, *19*(18), 11388. https://doi.org/10.3390/ijerph191811388

12 Current Challenges and Future Perspectives

12.1 SYSTEMIC BARRIERS AND HOW TO OVERCOME THEM

12.1.1 Institutional and Disciplinary Fragmentation

One of the most persistent systemic barriers to effective environmental education is the institutional fragmentation that characterizes contemporary education systems. This fragmentation manifests itself at multiple levels, from the disciplinary organization of curricula to the separation of formal, nonformal, and informal education. Fortunato et al. (2025) highlight that the intercultural approach to environmental education, in conjunction with the 2030 Agenda, underscores the challenges educators encounter in integrating intercultural and environmental practices within the confines of rigid educational frameworks and institutional constraints.

Overcoming this fragmentation requires systemic approaches that integrate different dimensions of human knowledge and experience. Educational institutions must evolve toward more flexible models that allow transdisciplinary collaboration and the development of integrated competencies. This implies not only changes in curriculum organization but also transformations in organizational cultures that have traditionally privileged disciplinary specialization over holistic integration.

The study conducted by Acuña-Rodríguez et al. (2023) presents a systematic review of the integration of environmental education within the curricula of higher education institutions. The findings indicate that numerous universities are opting to incorporate the transversality of environmental education across all professional disciplines, emphasizing comprehensive training and recognizing the potential contributions of individuals toward achieving a sustainable and ecological nation. This cross-cutting approach represents a promising strategy for overcoming traditional disciplinary fragmentation.

12.1.2 Political and Economic Resistance

Political and economic barriers to effective environmental education reflect the broader tensions between short-term economic interests and long-term sustainability imperatives. In her 2015 study on neoliberalism, pluralism, and environmental education, Kopnina (2015) elucidates that neoliberalism promotes economic growth through open markets while often neglecting sustainability imperatives.

These resistances manifest in multiple ways, including inadequate funding for environmental education programs, policies that prioritize quantifiable educational outcomes over qualitative transformations, and pressures to align education with immediate economic demands rather than long-term sustainability needs. Overcoming these barriers requires multifaceted strategies, including political advocacy, developing

DOI: 10.1201/9781003779667-16

evidence of the economic benefits of environmental education, and building broad coalitions to influence decision-making processes.

The need for clarity and specificity in environmental education policy design is well documented across diverse national contexts. Kopnina (2020) argued that environmental education policies frequently suffer from vagueness in their objectives and implementation guidelines, creating significant gaps between policy intentions and classroom practice. Her critical evaluation found that without practical and structured policy frameworks that clearly articulate expectations, provide concrete implementation guidance, and establish accountability mechanisms, environmental education remains marginalized within broader educational systems, regardless of the level of political rhetoric supporting it.

12.1.3 Capacity and Resource Constraints

Capacity and resource constraints represent operational barriers that can impede the effective implementation of environmental education programs, even when political and social support exists. These constraints include deficits in educator training, inadequate infrastructure, insufficient educational materials, and limited technical support.

Gebrekidan and Gebremedhin (2024) examine the integration and efficacy of formal environmental education in Africa and India, highlighting that "due to limited governmental prioritization, widespread integration challenges, the absence of a comprehensive strategy, and financial and technological limitations, environmental education in Africa remains unimplemented, predominantly field-based, and ineffective."

Overcoming these limitations requires systematic investments in capacity building that encompass several dimensions. This includes comprehensive teacher-training programs that integrate technical and pedagogical skills, infrastructure development to support innovative educational methodologies, and the creation of technical support systems that provide ongoing assistance to educators and institutions.

12.1.4 Disconnect between Research and Practice

A significant barrier to progress in this field is the persistent disconnect between academic research and educational practice. This disconnect manifests as a failure to translate research findings into practical methodologies, the development of assessment systems that do not reflect the complexity of environmental education processes, and limited opportunities for feedback between researchers and practitioners.

Ardoin et al. (2020) conducted a systematic review of environmental education outcomes for conservation and identified multiple evidence-based principles for improving program effectiveness across diverse delivery modalities, including online and hybrid formats. However, their analysis also revealed a persistent gap between research evidence and classroom implementation, noting that many effective principles identified through rigorous research remain underutilized in practice. Implementing these principles requires strengthening communication between researchers and educators through clearer, more effective channels that facilitate knowledge transfer and the development of applicable tools.

Overcoming this disconnect requires the development of more effective interfaces between research and practice, including collaborative research programs, dissemination systems that make research findings accessible to practitioners, and evaluation frameworks that generate practically useful evidence for program improvement.

12.1.5 Integrated Strategies for Overcoming

Overcoming systemic barriers involves applying comprehensive strategies that simultaneously address multiple dimensions of the problem. These strategies should include components of institutional transformation, capacity building, political advocacy, and the construction of social movements capable of generating pressure for systemic changes.

A crucial element is developing collaborative governance frameworks that can align efforts across different sectors and organizational levels. Velempini and Maruatona (2025) highlight UNESCO's impact on the delivery of continuing and environmental education in Botswana, noting that "although Faure's report did not appear to have been formally endorsed, Botswana has benefited directly and indirectly from UNESCO's technical assistance over time." This observation underscores the significance of international frameworks in providing guidance and support for national development.

12.2 EMERGING TRENDS: ARTIFICIAL INTELLIGENCE (AI), VIRTUAL REALITY, HYBRID EDUCATION

12.2.1 AI and Educational Personalization

The integration of AI into environmental education presents unprecedented opportunities to personalize learning experiences, automate assessment processes, and provide adaptive feedback, thereby significantly improving educational effectiveness. AI systems can analyze individual learning patterns to tailor content, pedagogical sequences, and methodologies to meet each student's specific needs.

Research on the effectiveness of digital media in environmental education reveals significant variability in outcomes depending on the design and delivery approach employed. Rousell and Cutter-Mackenzie-Knowles (2020), in their systematic review of climate change education, found that while digital technologies offer significant potential for expanding the reach and engagement of environmental education, their effective implementation requires careful attention to pedagogical design, audience characteristics, and engagement factors. Their analysis demonstrated that digital platforms focused primarily on information transmission consistently underperformed compared to those incorporating interactive, participatory, and emotionally engaging elements, suggesting that the technological medium alone is insufficient without thoughtful pedagogical integration.

The most promising AI systems for environmental education include intelligent tutors that adapt instruction to individual progress, automated assessment systems that provide immediate feedback on environmental competencies, and recommendation platforms that suggest resources and activities based on the specific interests and needs of the students. However, effective implementation requires careful consideration of ethical issues, including data privacy, algorithmic bias, and equity in access to advanced technologies.

12.2.2 Virtual Reality and Immersive Experiences

Virtual and augmented reality technologies are transforming the possibilities for creating immersive environmental educational experiences that can transport students to remote ecosystems, simulate complex environmental processes, and provide opportunities for safe experimentation with diverse scenarios.

Petersen et al. (2020) investigated how to optimize immersive virtual learning in climate change education and demonstrated that carefully designed virtual and augmented reality experiences can create meaningful and memorable learning outcomes. Their experimental study found that immersive technologies significantly enhance students' knowledge acquisition, emotional engagement, and pro-environmental attitudes when they incorporate evidence-based multimedia learning principles, including embodied interaction, guided exploration, and opportunities for reflective processing. These findings suggest that the fusion of immersive technologies with sound pedagogical design has the potential to create engaging and transformative learning experiences that promote environmental awareness and responsibility.

The most promising applications of virtual reality in environmental education include ecosystem simulations that enable detailed exploration of natural environments, environmental impact modeling that visualizes the consequences of different decisions, and multiperspective experiences that allow students to view environmental systems from the perspectives of different species or stakeholders.

12.2.3 Hybrid Education Models

The convergence of face-to-face and digital education has given rise to hybrid models that combine the strengths of different educational modalities. These models can provide flexibility for diverse students while maintaining the experiential elements crucial to effective environmental education.

Punzalan (2024), in an analysis of the integration of environmental education into science e-learning during the COVID-19 pandemic, found that "teachers successfully integrated environmental education into their e-learning practices, often focusing on real-life environmental scenarios." Qualitative analysis revealed themes including "how environmental education is integrated into science e-learning, teaching strategies and resources used for environmental education integration, challenges faced in this process, and the emergence of best practices for effective environmental education integration."

The most effective hybrid models seamlessly integrate physical and digital experiences, using technology to amplify rather than replace direct interactions with natural systems. This can include the use of sensors for real-time environmental monitoring, digital platforms for data analysis and sharing, and communication tools to connect students with experts and global communities to promote environmental awareness.

12.2.4 Challenges of Technology Implementation

The effective implementation of emerging technologies in environmental education faces multiple challenges, including barriers to access, teacher training needs, environmental sustainability concerns, and the risks of technological dependence that can diminish direct connections with nature.

Wamsler et al. (2021) emphasized that the relationship between technology and environmental education must be critically examined to ensure that digital tools serve transformative educational objectives rather than merely reproducing conventional pedagogical approaches in digital formats. Their analysis found that the commitment of educational institutions and the development of social incentives to increase environmental awareness and promote ecological behaviors are critical to successful technology implementation. The authors concluded that technological integration is most effective when it is embedded within broader institutional strategies that align digital innovation with pedagogical goals, community engagement, and sustainability values.

12.2.5 Innovations in Assessment and Feedback

Emerging technologies are creating new possibilities for assessment and feedback in environmental education that can capture aspects of learning that have been difficult to measure. Olsson et al. (2022), in their longitudinal study of secondary school students' sustainability consciousness, demonstrated that effective assessment systems must capture changes across multiple dimensions of learning—including cognitive understanding, values orientation, and behavioral intentions—over extended time periods. Their research provides methodological insights into how continuous assessment frameworks can track the progressive development of environmental competencies, offering a model for the kind of multidimensional evaluation that emerging technologies can facilitate at scale. This includes continuous assessment systems, behavioral analysis in virtual environments, and digital portfolio tools that can document long-term progress.

The most innovative assessment systems combine automated analysis and human review to provide comprehensive feedback, encompassing the cognitive, affective, and behavioral dimensions of environmental learning. These systems can identify patterns of engagement, analyze the quality of written reflections, and provide personalized recommendations for continuous improvement in the learning process.

12.3 ENVIRONMENTAL EDUCATION IN THE ANTHROPOCENE ERA

12.3.1 Paradigmatic Reconceptualization

The Anthropocene requires a fundamental reconceptualization of environmental education to acknowledge that humans have become the dominant geological force, permanently altering planetary systems. This reconceptualization must move beyond traditional educational frameworks that assume a separation between society and nature and embrace approaches that recognize the complex integration of human and natural systems.

Reid et al. (2021) emphasize the critical role of environmental education in addressing scientific warnings and advocate for its reimagining, recreation, and restoration. They assert that to prevent endangering both our future and that of the planet and to facilitate a just transition to sustainability, environmental education serves as a fundamental component for the anticipated social and environmental transformations highlighted in such warnings. The authors emphasize that "consensus on our environmental predicaments is not simply a matter for scientists; it must be supported in multiple spheres."

This reconceptualization must include the development of new competencies that can prepare students to navigate the uncertainty, complexity, and accelerated change that characterize the Anthropocene epoch. This includes competencies in systems thinking, adaptability, and transformational leadership, which may be necessary for building resilient societies.

12.3.2 Ecocentric and Transformational Approaches

Environmental education in the Anthropocene must embrace ecocentric approaches that decenter dominant human perspectives to develop a deeper understanding of ecological interdependencies. The research by Molina-Motos (2019), which articulates ecophilosophical principles for ecocentric environmental education, proposes "the establishment of some principles derived from genuinely ecological and ecocentric environmental philosophies."

The study reviews "the most significant contributions of earth ethics, deep ecology, social ecology, ecofeminism, and paradigm shift ecologies" to develop "a set of facets, key categories, and characteristics that offer an integrated and synoptic view of how Ecocentric Environmental Education could be based on eco-physical principles." This approach represents a promising direction for environmental education that can respond to the realities of the Anthropocene era.

Transformational approaches must include environmental justice, recognition of the rights of nature, and the development of new forms of citizenship that extend beyond human communities to encompass broader ecological systems that include nonhuman beings.

12.3.3 Pedagogies of Urgency and Hope

Environmental education in the Anthropocene must balance the recognition of the urgency of environmental crises with the cultivation of hope and agency that can motivate effective action. This balance requires sophisticated pedagogies that convey the seriousness of environmental challenges without inducing feelings of paralysis or despair.

Reid et al. (2021), in their analysis of the need to reimagine, recreate, and restore environmental education, explore the opportunities to effect the changes in attitude, values, and behaviors needed for collective survival in the context of escalating environmental crises. The authors emphasize the need for foresight in educational and policy planning to address the magnitude of problems that threaten life on this planet. Their work underscores that effective pedagogies of urgency and hope must balance rigorous scientific communication about environmental crises with the cultivation of agency, resilience, and constructive engagement among learners.

The most effective pedagogies integrate rigorous scientific information about environmental crises with narratives of successful solutions, models of inspiring leadership, and opportunities for meaningful participation in actions to effect change. This includes connections with environmental social movements, ecological restoration projects, and sustainable innovation initiatives.

12.3.4 Integration of Diverse Knowledge and Epistemologies

Environmental education for the Anthropocene must integrate diverse epistemologies, including Western scientific knowledge, traditional indigenous knowledge, and emerging forms of knowledge arising from interdisciplinary syntheses. This integration is crucial for developing comprehensive and effective responses to complex environmental challenges.

Fortunato et al. (2025) highlight the significance of conducting a more critical analysis of sustainability within broader socioeconomic frameworks and underscore the essential role of educators, who must navigate the intersection of societal expectations and the realities they encounter amidst global challenges.

Responding to this challenge requires educators to cultivate intercultural competencies, underpinned by curricular frameworks that accommodate diverse ways of knowing and pedagogical methodologies that foster respectful dialogue across epistemological traditions.

12.3.5 Ethical and Spiritual Dimensions

Environmental education in the Anthropocene must address the ethical and spiritual dimensions of human–nature relationships, which may be fundamental to profound transformations in value and behavior. This includes exploring different ethical frameworks for environmental decision-making, developing spiritual connections with natural systems, and cultivating virtues that support sustainable ways of life.

Cholil and Parker (2021) explore the implementation of environmental education and eco-theology within Franciscan schools in Indonesia. Their study examines the practical application of St. Francis of Assisi's eco-theology and assesses the extent to which students comprehend the environmental implications outlined in the Pope's Encyclical, Laudato Si'. The findings underscore the significance of St. Francis' eco-theology in the context of global efforts toward environmental sustainability. Ethical and spiritual dimensions can provide deep motivation for sustained environmental engagement. They may be especially important for developing the emotional resilience required to confront the scale of contemporary environmental challenges.

12.4 VISION 2030–2050: TOWARD PLANETARY CITIZENSHIP

12.4.1 Agenda 2030 and Educational Transformation

The 2030 Agenda for Sustainable Development provides a global framework that significantly influences the evolution of environmental education toward more integrated and action-oriented approaches. Research conducted by Ariza et al. (2025) on environmental education, bioethics, and the 2030 Agenda as an urgent episteme in universities posits that "the level of achievement of the agenda is not keeping pace with what is set out in the Sustainable Development Goals." The study further underscores "the role of universities in the coming years in generating an environmental culture that contributes to the goals of the 2030 Agenda."

The educational transformation required to meet the goals of the 2030 Agenda includes integrating environmental education with global citizenship education, developing skills for international collaboration, and cultivating a systemic understanding of the interconnections between social, economic and environmental challenges.

Ali et al. (2025), in their research on enhancing environmental education concerning mangroves and climate change to fulfill Sustainable Development Goal (SDG) 13 in coastal communities, reveal that "enhancing environmental education through participatory methods and the integration of traditional ecological knowledge with scientific frameworks can enhance community engagement and policy alignment." This approach illustrates the potential of environmental education to contribute directly to the achievement of specific development goals.

12.4.2 Concept of Planetary Citizenship

Planetary citizenship is emerging as a central concept in environmental education, recognizing that environmental challenges transcend national borders and require identities and responsibilities that extend to global and planetary scales. Brundiers et al. (2021), in their agreed-upon reference framework for key competencies in sustainability education, identified several competencies that are foundational to planetary citizenship, including systems thinking at global scales, anticipatory competence for understanding long-term planetary trajectories, and normative competence for navigating diverse cultural perspectives on environmental responsibility. This conceptualization entails developing competencies in understanding planetary systems, a commitment to intergenerational equity, and participating in global governance.

Planetary citizenship encompasses cognitive (understanding planetary processes), affective (emotional connection to global systems), and behavioral (participation in actions with a planetary impact) dimensions. The development of these dimensions requires educational approaches that connect local experiences with global processes and provide opportunities for meaningful participation in planetary-scale actions.

Severino-Gonzalez et al. (2024) examine university social responsibility and environmental education, focusing on the challenges posed by higher education from the perspective of institutional policies as a means of articulating strategies that contribute to the development of socially responsible professionals. Their analysis underscores the significance of implementing educational innovations that address the tensions inherent in modernity.

12.4.3 21st-Century Transformational Competencies

The competencies required for effective planetary citizenship include cognitive, socio-emotional, and behavioral capacities that enable individuals to contribute to the systemic transformations necessary for planetary sustainability. These competencies include systemic thinking, intercultural collaboration, adaptive leadership, and sustainable innovation.

Systems thinking includes the ability to understand complex interconnections, anticipate unintended consequences, and identify leverage points for systemic changes. Intercultural collaboration requires competencies to work effectively with people from different cultural backgrounds, communicate across linguistic and conceptual differences, and build a consensus around shared goals.

Innovative approaches to developing emotional engagement with environmental issues are increasingly recognized as essential complements to cognitive and behavioral strategies. Chawla (2020), in her comprehensive review of childhood nature connection, demonstrated that emotional engagement—including experiences of joy, wonder, and even humor—plays a fundamental role in developing lasting connections with natural systems. Her analysis found that educational approaches incorporating positive emotional experiences alongside scientific content consistently produced stronger and more durable pro-environmental attitudes and behaviors, suggesting that creative pedagogical strategies that cultivate positive emotional connections with nature can serve as powerful tools for environmental education across diverse cultural contexts.

12.4.4 Technologies for Global Scaling

Fraisl et al. (2022), in their comprehensive review of citizen science in environmental and ecological sciences, demonstrated that digital platforms enabling global participation in environmental data collection and analysis represent one of the most successful models for scaling environmental education beyond traditional institutional boundaries. Their analysis found that citizen-science platforms simultaneously advance scientific research and provide transformative learning experiences, connecting participants across geographic and cultural boundaries through shared environmental inquiry. The development of planetary citizenship requires technologies that facilitate global collaboration, enable knowledge sharing across borders, and provide equitable access to high-quality educational opportunities worldwide. These technologies include global learning platforms, machine translation systems, and virtual collaboration tools that connect students and educators across the globe.

The most promising technologies for global scaling include massive open online course (MOOC) platforms specifically designed for environmental education, educational social networks that facilitate global exchange among environmental educators, and gamification systems that motivate participation in environmental actions on a global scale.

de Andrade (2024), in a study of environmental education in the Brazilian Amazon from an environmental governance perspective, found that "the social representations of the environment and environmental education developed by teachers provided powerful elements for the defense of nature aimed at environmental awareness and citizen participation in decision-making processes." This evidence suggests that locally rooted approaches can contribute to the global goals of planetary citizenship.

12.4.5 Future Scenarios and Adaptive Planning

Developing a vision for 2030–2050 requires consideration of multiple future scenarios that reflect different possible trajectories for global socioecological systems. Menton et al. (2020), in their analysis of environmental justice and the SDGs, demonstrated that future scenarios must account for the uneven distribution of environmental risks and educational opportunities across different communities, emphasizing that adaptive planning frameworks should explicitly incorporate equity considerations to ensure that environmental education responses do not exacerbate existing inequalities. These scenarios may include optimistic futures in which environmental education contributes to successful transformations toward sustainability, as well as more challenging scenarios in which accelerated environmental crises require emergency educational responses to mitigate the damage.

Adaptive planning for these scenarios requires the development of organizational and institutional capacities to respond flexibly to changes in conditions and opportunities. This includes monitoring systems that can identify emerging changes, decision-making frameworks that can quickly evaluate response options, and organizational cultures that value experimentation and continuous learning.

Fraisl et al. (2022) further emphasize that environmental education requires various forms of integrated, technology-enhanced, and project-based teaching and learning approaches that involve hands-on experimentation along with the retrieval and critical analysis of information from diverse sources and perspectives. This perspective underscores the importance of methodological flexibility in responding to an uncertain future.

12.5 CONCLUSION

12.5.1 Summary of Necessary Transformations

The analysis of current challenges and prospects for environmental education reveals the need for fundamental transformations that go beyond incremental improvement to embrace a systemic reimagining of the field. These transformations must simultaneously address structural barriers, leverage technological opportunities, respond to the realities of the Anthropocene, and build the capacity for planetary citizenship.

The systemic barriers identified—including institutional fragmentation, political and economic resistance, capacity constraints, and disconnects between research and practice—require coordinated strategies to effect change at multiple levels simultaneously. Effectively overcoming these barriers demands not only technical interventions but also transformations in the values, power structures, and conceptual frameworks that guide the education systems.

The evidence presented indicates that while significant challenges exist, extraordinary opportunities for transformation are emerging. Emerging technologies offer unprecedented possibilities for educational personalization, immersive experiences, and global scalability. However, their effective implementation requires careful attention to equity, sustainability, and alignment with fundamental educational values.

12.5.2 Imperatives for Immediate Action

The urgency of contemporary environmental crises creates imperatives for immediate action that cannot wait for the complete resolution of systemic challenges to sustainability. These imperatives include rapidly scaling effective educational programs, integrating environmental education into climate emergency responses, and mobilizing resources to accelerate this educational transformation.

Whitburn et al. (2020), in their meta-analysis of human connection to nature and pro-environmental behavior, demonstrated that environmental education serves as a powerful tool for creating knowledge, awareness, and positive attitudes that encourage active participation in environmental protection across diverse populations, including rural communities. Their analysis confirmed that educational interventions emphasizing direct connections with nature and locally relevant environmental challenges are particularly effective in motivating sustained behavioral change, suggesting that environmental education can immediately contribute to climate adaptation and mitigation efforts even in resource-constrained contexts.

The imperatives for immediate action also include developing crisis response capacities, such as environmental education in disaster preparedness strategies, early warning systems, and post-crisis recovery programs. These capacities require educational frameworks that can operate effectively under stressful and uncertain conditions.

12.5.3 Directions for Future Research

Priority areas for future research include developing more robust theoretical frameworks for environmental education in the Anthropocene, evaluating the effectiveness of emerging technologies in diverse educational contexts, and analyzing the factors that facilitate systemic transformation in educational institutions.

Future research should also address equity and justice in access to high-quality environmental education, develop methodologies for assessing long-term impacts, and explore intersections between environmental education and other social transformation movements.

Redman et al. (2018), in their analysis of continuing professional development in sustainability education for K-12 teachers, identified key principles and programmatic approaches for strengthening educators' capacity to deliver effective environmental education. Their review found that despite growing availability of professional development programs, many teachers face institutional, curricular, and professional barriers that constrain their capacity to translate sustainability commitments into sustained classroom practice. This perspective underscores the need for research that directly informs professional development programs and provides practical support for educators.

12.5.4 Call for Transformative Action

The evidence presented constitutes an urgent call for transformative action that can reposition environmental education as a central force in building a sustainable future. This requires commitment from multiple sectors, including governments, educational institutions, civil society organizations, the private sector, and local communities.

Transformative action must be guided by the principles of equity, inclusion, and justice, which ensure that the benefits of improved environmental education are accessible to all communities, particularly those that have been historically marginalized. It must also be characterized by an appropriate sense of urgency that recognizes the magnitude of environmental crises while maintaining a commitment to quality education and profound transformation.

Kalsoom et al. (2022) present a collaborative reflection on environmental practices as a means of advancing environmental education within teacher training. Their findings indicate that engaging in collaborative reflection on environmental issues and practices fosters pro-environmental behaviors among participants. This evidence substantiates the efficacy of approaches that prioritize collaboration and reflection as fundamental components of educational transformation.

12.5.5 Vision for the Next Decade

The vision for the next decade includes environmental education that is universally accessible, culturally relevant, technologically sophisticated, and transformatively effective. This vision requires the construction of educational systems that can respond quickly to changes in environmental conditions while maintaining a commitment to human development and social justice. Olsson et al. (2022) demonstrated through longitudinal evidence that sustained educational engagement produces cumulative effects on sustainability consciousness, confirming that the long-term investments envisioned here can yield transformative outcomes when systematically implemented.

The realization of this vision depends on sustained investments in human and institutional capacities, the development of appropriate technologies, and the building of social movements capable of sustaining the political and social commitment necessary for systemic transformation. It also requires cultivating hope, creativity, and determination to sustain actions over the decades required to achieve profound transformations.

Finally, future environmental education must prepare individuals and communities not only to adapt to environmental change but also to lead the construction of more just, sustainable, and resilient worlds. This preparation requires sophisticated combinations of knowledge, skills, values, and capacities for action that can empower people to become effective agents of transformation for planetary sustainability.

REFERENCES

Acuña-Rodríguez, M., Rincón, L. C., González, A. I., López, Y. G., Umaña-Ibáñez, S. F., Coronado-Hernández, J. R., & Gatica, G. (2023). Articulation of Environmental Education in the Curricula of Higher Education Institutions: A Systematic Review. Perspectives and Trends in Education and Technology, Singapore.

Ali, M. I., Malik, A., Aslinda, A., Umar, R., & Dg Bau, Q. (2025). Enhancing environmental education on Mangroves and climate change to achieve SDG 13 in coastal communities of South Sulawesi, Indonesia. *International Research in Geographical and Environmental Education*, 1–20. https://doi.org/10.1080/10382046.2025.2524673

Ardoin, N. M., Bowers, A. W., & Gaillard, E. (2020). Environmental education outcomes for conservation: A systematic review. *Biological Conservation*, *241*, 108224. https://doi.org/10.1016/j.biocon.2019.108224

Ariza, C. P., Zubiria, L. M. A., & Daza, B. C. O. (2025, enero–abril). Educación ambiental, bioética y Agenda 2030: un episteme urgente en la universidad. *Encuentros. Revista de Ciencias Humanas, Teoría Social y Pensamiento Crítico*, (23), 174–184. https://doi.org/10.5281/zenodo.14268438

Brundiers, K., Barth, M., Cebrián, G., Cohen, M., Diaz, L., Doucette-Remington, S., Dripps, W., Habron, G., Harré, N., Jarchow, M., Losch, K., Michel, J., Mochizuki, Y., Rieckmann, M., Parnell, R., Walker, P., & Zint, M. (2021). Key competencies in sustainability in higher education—toward an agreed-upon reference framework. *Sustainability Science*, *16*(1), 13–29. https://doi.org/10.1007/s11625-020-00838-2

Chawla, L. (2020). Childhood nature connection and constructive hope: A review of research on connecting with nature and coping with environmental loss. *People and Nature*, *2*(3), 619–642. https://doi.org/10.1002/pan3.10128

Cholil, S., & Parker, L. (2021). Environmental education and eco-theology: Insights from Franciscan schools in Indonesia. *Environmental Education Research*, *27*(12), 1759–1782. https://doi.org/10.1080/13504622.2021.1968349

de Andrade, F. M. R. (2024). Environmental Education in the Brazilian Amazon, in Pará State: The meanings of environmental governance. *Asia Pacific Education Review*, *25*(4), 871–883. https://doi.org/10.1007/s12564-023-09841-5

Fortunato, I., Mena, J., & Vatter, C. (2025). From an intercultural approach to environmental education to 2030 Agenda: Exploring teachers' role. *Intercultural Education*, 1–17. https://doi.org/10.1080/14675986.2025.2535074

Fraisl, D., Hager, G., Bedessem, B., Gold, M., Hsing, P.-Y., Danielsen, F., Hitchcock, C. B., Hulbert, J. M., Piera, J., Spiers, H., Thiel, M., & Haklay, M. (2022). Citizen science in environmental and ecological sciences. *Nature Reviews Methods Primers*, *2*(1), 64. https://doi.org/10.1038/s43586-022-00144-4

Gebrekidan, T., & Gebremedhin, G. (2024). Integration and effectiveness of formal environmental education in Africa and India: Review. *European Journal of Sustainable Development Research*, *8*(2), em0256. https://doi.org/10.29333/ejosdr/14368

Kalsoom, Q., Khanam, A., & Qureshi, N. (2022). Collaborative reflection on environmental practices: A vehicle for environmental education in teacher education. *Reflective Practice*, *23*(2), 162–176. https://doi.org/10.1080/14623943.2021.2001320

Kopnina, H. (2015). Neoliberalism, pluralism and environmental education: The call for radical re-orientation. *Environmental Development*, *15*, 120–130. https://doi.org/10.1016/j.envdev.2015.03.005

Kopnina, H. (2020). Education for the future? Critical evaluation of education for sustainable development goals. *The Journal of Environmental Education*, *51*(4), 280–291. https://doi.org/10.1080/00958964.2019.1710444

Menton, M., Larrea, C., Latorre, S., Martinez-Alier, J., Peck, M., Temper, L., & Walter, M. (2020). Environmental justice and the SDGs: From synergies to gaps and contradictions. *Sustainability Science*, *15*(6), 1621–1636. https://doi.org/10.1007/s11625-020-00789-8

Molina-Motos, D. (2019). Ecophilosophical principles for an ecocentric environmental education. *Education Sciences*, *9*(1), 37. https://doi.org/10.3390/educsci9010037

Olsson, D., Gericke, N., & Boeve-de Pauw, J. (2022). The effectiveness of education for sustainable development revisited: A longitudinal study on secondary students' action competence for sustainability. *Environmental Education Research*, *28*(3), 405–429. https://doi.org/10.1080/13504622.2022.2033170

Petersen, G. B., Klingenberg, S., Mayer, R. E., & Makransky, G. (2020). The virtual field trip: Investigating how to optimize immersive virtual learning in climate change education. *British Journal of Educational Technology*, *51*(6), 2099–2115. https://doi.org/10.1111/bjet.12991

Punzalan, C. (2024). Environmental education in junior high school science: Teachers' integration perceptions through distance learning modality. *Jurnal Pendidikan IPA Indonesia*, *13*(2), 358–366. https://doi.org/10.15294/jjqkgj45

Redman, E., Wiek, A., & Redman, A. (2018). Continuing professional development in sustainability education for K-12 teachers: Principles, programme, applications, outlook. *Journal of Education for Sustainable Development*, *12*(1), 59–80. https://doi.org/10.1177/2455133318777182

Reid, A., Dillon, J., Ardoin, N., & Ferreira, J.-A. (2021). Scientists' warnings and the need to reimagine, recreate, and restore environmental education. *Environmental Education Research*, *27*(6), 783–795. https://doi.org/10.1080/13504622.2021.1937577

Rousell, D., & Cutter-Mackenzie-Knowles, A. (2020). A systematic review of climate change education: Giving children and young people a 'voice' and a 'hand' in redressing climate change. *Children's Geographies*, *18*(2), 191–208. https://doi.org/10.1080/14733285.2019.1614532

Severino-Gonzalez, P., Gallardo-Vázquez, D., Saldía-Barahona, H., Villanueva-Arequipeño, T., Sarmiento-Peralta, G., & Romero-Argueta, J. D. J. (2024). University social responsibility and environmental education: Challenges for the training of socially responsible professionals. *Interciencia*, *49*(2), 104–110.

Velempini, K., & Maruatona, T. (2025). A critical review of UNESCO's influence on lifelong learning and environmental education delivery in Botswana. *International Journal of Lifelong Education*, 1–22. https://doi.org/10.1080/02601370.2025.2497411

Wamsler, C., Osberg, G., Osika, W., Herndersson, H., & Mundaca, L. (2021). Linking internal and external transformation for sustainability and climate action: Towards a new research and policy agenda. *Global Environmental Change*, *71*, 102373. https://doi.org/10.1016/j.gloenvcha.2021.102373

Whitburn, J., Linklater, W., & Abrahamse, W. (2020). Meta-analysis of human connection to nature and proenvironmental behavior. *Conservation Biology*, *34*(1), 180–193. https://doi.org/10.1111/cobi.13381

Index

P

Q

R

S

T

U

V

W

Y

For Product Safety Concerns and Information please contact our EU representative GPSR@taylorandfrancis.com Taylor & Francis Verlag GmbH, Kaufingerstraße 24, 80331 München, Germany

Batch number: 10725802

Printed by Printforce, the Netherlands